General View of the Agriculture of the
County of Clydesdale

British Statute Miles.

General View
of the
AGRICULTURE
of the
COUNTY
of
CLYDESDALE

with

OBSERVATIONS ON THE MEANS OF ITS IMPROVEMENT
DRAWN UP FOR THE CONSIDERATION OF

THE BOARD OF AGRICULTURE,

AND INTERNAL IMPROVEMENT

by

JOHN NAISMITH
1806

Glasgow
The Grimsay Press
2003

The Grimsay Press
an imprint of
Zeticula
57 St Vincent Crescent
Glasgow
G3 8NQ

http://www.thegrimsaypress.co.uk
admin@thegrimsaypress.co.uk

Transferred to digital printing in 2003

First published in Great Britain 1806 by Richard
Phillips, London

ISBN 0 902664 44 1

Reproduced from the copy in the Library of the Scottish
Agricultural College, Ayr Campus, Scotland

Contents

ADVERTISEMENT.

———

THE great desire that has been very generally expressed, for having the AGRICULTURAL SURVEYS of the KINGDOM reprinted, with the additional Communications which have been received since the ORIGINAL REPORTS were circulated, has induced the BOARD OF AGRICULTURE to come to a resolution of reprinting such as may appear on the whole fit for publication. It is proper at the same time to add, that the Board does not consider itself responsible for any fact or observation contained in the Reports thus reprinted, as it is impossible to consider them yet in a perfect state ; and that it will thankfully acknowledge any additional information which may still be communicated : an invitation, of which, it is hoped, many will avail themselves, as there is no circumstance from which any one can derive more real satisfaction, than that of contributing, by every possible means, to promote the improvement of his Country.

———

N. B. *Letters to the Board, may be addressed to Lord* SHEFFIELD, *the President, No. 32, Sackville-Street, Piccadilly, London.*

PLAN

PLAN

FOR RE-PRINTING THE

AGRICULTURAL SURVEYS.

BY THE PRESIDENT OF THE BOARD OF AGRICUL-
TURE.

A BOARD established for the purpose of making every essential inquiry into the Agricultural State, and the means of promoting the Internal Improvement of a powerful Empire, will necessarily have it in view to examine the sources of public prosperity, in regard to various important particulars. Perhaps the following is the most natural order for carrying on such important investigations; namely, to ascertain,

1. The riches to be obtained from the surface of the national territory.

2. The mineral or subterraneous treasures of which the country is possessed.

3. The wealth to be derived from its streams, rivers, canals, inland navigations, coasts, and fisheries ;— and

4. The means of promoting the improvement of the people, in regard to their health, industry, and morals, founded on a *statistical* survey, or a minute and careful inquiry into the actual state of every parochial district in the kingdom, and the circumstances of its inhabitants.

Under one or other of these heads, every point of real importance that can tend to promote the general happiness of a great nation, seems to be included.

Investigations of so extensive and so complicated a nature, must require, it is evident, a considerable space of time before they can be completed. Differing indeed in many respects from each other, it is better perhaps that they should be undertaken at different periods, and separately considered. Under that impression, the Board of Agriculture has hitherto directed its attention to the first point only, namely, the cultivation of the surface, and the resources to be derived from it.

That the facts essential for such an investigation might be collected with more celerity and advantage, a number of intelligent and respectable individuals were appointed, to furnish the Board with accounts of the state of husbandry, and the means of improving the different districts of the kingdom. The returns they sent were printed, and circulated by every means the Board of Agriculture could devise, in the districts to which they respectively related; and in consequence of that circulation, a great mass of additional valuable information has been obtained. For the purpose of communicating that information to the Public in general, but more especially to those Counties the most interested therein, the Board has resolved to re-print the Survey of each County, as soon as it seemed to be fit for publication; and, among several equally advanced, the Counties of Norfolk and Lancaster were pitched upon for the commencement of the proposed publication; it being thought most advisable to begin with one County on the Eastern, and another on the Western Coast of the island. When all these Surveys shall have been thus re-printed, it will be attended with little difficulty to draw up

an

an abstract of the whole (which will not probably exceed two or three volumes quarto) to be laid before His Majesty, and both Houses of Parliament; and afterwards, a General Report on the present state of the country, and the means of its improvement, may be systematically arranged, according to the various subjects connected with Agriculture. Thus every individual in the kingdom may have,

1. An account of the husbandry of his own particular county; or,

2. A general view of the agricultural state of the kingdom at large, according to the counties, or districts, into which it is divided; or,

3. An arranged system of information on agricultural subjects, whether accumulated by the Board since its establishment, or previously known:

And thus information respecting the state of the kingdom, and agricultural knowledge in general, will be attainable with every possible advantage.

In re-printing these Reports, it was judged necessary, that they should be drawn up according to one uniform model; and after fully considering the subject, the following form was pitched upon, as one that would include in it all the particulars which it was necessary to notice in an Agricultural Survey. As the other Reports will be reprinted in the same manner, the reader will thus be enabled to find out at once where any point is treated of, to which he may wish to direct his attention.

PLAN OF THE RE-PRINTED REPORTS.

CHAP. VII. *Arable Land.*

CHAP. VIII. *Grass.*

CHAP. IX. *Gardens and Orchards.*

CHAP. X. *Woods and Plantations.*

CHAP. XI. *Wastes.*

* Where the quantity is considerable, the information respecting the crops commonly cultivated may be arranged under the following heads—for example, WHEAT:

1. Preparation { tillage, manure. }
2. Sort.
3. Steeping.
4. Seed (quantity sown).
5. Time of sowing.

6. Culture whilst growing { hoe, weeding, feeding. }
7. Harvest.
8. Thrashing.
9. Produce.
10. Manufacture of bread.

In general, the same heads will suit the following grains:
Barley. Oats. Beans. Rye. Pease. Buck-wheat.

Vetches — Application.

Cole-seed — { Feeding, Seed. }

Turnips — { Drawn, Fed, Fed on grass, —— in houses. }

CHAP.

PERFECTION in such inquiries is not in the power of any body of men to obtain at once, whatever may be the extent of their views or the vigour of their exertions. If LOUIS XIV. eager to have his kingdom known, and possessed of boundless power to effect it, failed so much in the attempt, that of all the provinces in his kingdom, only one was so described as to secure the approbation of posterity *, it will not be thought strange that a Board, possessed

* See VOLTAIRE's Age of LOUIS XIV. vol. ii. p. 127, 128, edit. 1752. The following extract from that work will explain the circumstance above alluded to :

" LOUIS had no COLBERT, nor LOUVOIS, when, about the year 1698, for the instruction of the Duke of BURGUNDY, he ordered each of the intendants to draw up a particular description of his province. By this means an exact account of the kingdom might have been obtained, and a just enumeration of the inhabitants. It was an useful work, though all the intendants had not the capacity and attention of Monsieur DE LAMOIGNON DE BAVILLE. Had what the King directed been as well executed, in regard to

every

sessed of means so extremely limited, should find it difficult
to reach even that degree of perfection which perhaps
might have been attainable with more extensive powers.
The candid reader cannot expect in these Reports more
than a certain portion of useful information, so arranged as
to render them a basis for further and more detailed inquiries. The attention of the intelligent cultivators of the
kingdom, however, will doubtless be excited, and the minds
of men in general gradually brought to consider favoura-
bly of an undertaking which will enable all to contribute
to the national stores of knowledge, upon topics so truly
interesting as those which concern the agricultural interests of their country ; interests which, on just principles,
never can be improved, until the present state of the kingdom is fully known, and the means of its future improvement ascertained with minuteness and accuracy.

every province, as it was by this magistrate in the account of Languedoc, the
collection would have been one of the most valuable monuments of the age.
Some of them are well done ; but the plan was irregular and imperfect, because
all the intendants were not restrained to one and the same. It were to be
wished that each of them had given, in columns, the number of inhabitants
in each election ; the nobles, the citizens, the labourers, the artisans, the
mechanics ; the cattle of every kind ; the good, the indifferent, and the bad
lands ; all the clergy, regular and secular ; their revenues, those of the towns,
and those of the communities.

" All these heads, in most of their accounts, are confused and imperfect ;
and it is frequently necessary to search with great care and pains, to find what
is wanted. The design was excellent, and would have been of the greatest use,
had it been executed with judgment and uniformity."

PREFACE.

THE substance of this General View of the Agriculture, &c. of Clydesdale, was first printed and distributed in the beginning of the year 1794. It was afterwards put under a different arrangement, and, with various additions and improvements, was again published in 1798. That impression being now sold off, the booksellers have for some time been soliciting the Board of Agriculture for a new edition. The Honourable Board have therefore remitted to the original Writer to revise the Work, and make such corrections and improvements as shall appear requisite. The public have been for seven years in possession of the Work, and, in that time, it has been perused by many people of different descriptions in the county. The Writer has frequently conversed with a number of those, and has had occasion to visit almost all the different districts; but has found no cause to retract any account he has given, or to suspect that the description of things, as they then stood, was deficient; so that he finds little occasion for correction. Some alterations, however, are made on such articles as seemed to admit of improvement; but with

whatever

whatever fidelity the portrait may have been
drawn at the time, the change of circumstances
which have occurred, during the eventful period
which has since elapsed, is great, and deserves
to be recorded.   It is proposed, therefore, to
reprint the work as in the last impression, with
any additional notes and remarks which may
occur; and where any material alteration has
taken place, to contrast the circumstances of one
period with those of the other.

The Writer of this having taken leave of the
public at the last publication, it may be said, he
should not have again appeared ; but he did not
expect to live to see the changes which have
occurred, nor a new edition demanded.   Now
that it is demanded, he is, perhaps, somewhat
in the spirit of a certain gentleman, who being
about to lay out the grounds around his house,
was exhorted by his friends to call the assistance
of a professional man ; to which he answered,
that as he was to do it for his own satisfaction,
he was determined the work should be directed
solely by himself ; and he would rather that his
execution should be found fault with, than that
of another obtain praise.   Besides, as he still
entertains the same high idea of the importance
of Agriculture, and the inseparable connexion
which exists between its success and the public
prosperity, he has of late years employed himself
in examining if any means can be found, by
                                        which

which that Art may be delivered from the dark dominion of empiricism, and established upon plain and regular principles, applicable to all purposes. As his sentiments on the subject may possibly never appear in public, he has taken the opportunity, in the course of this Work, to introduce a few hints, in which he differs from the ordinary practice and opinion of Agriculturists, that those may excite inquiry, and lead to the discovery of the truth, and thus his private studies tend to the public benefit.

AGRICULTURAL SURVEY

OF

CLYDESDALE.

═══════════

CHAP. I.

GEOGRAPHICAL STATE AND CIRCUMSTANCES.

────◆────

SECT. I.—SITUATION AND EXTENT.

CLYDESDALE is so called from the noble river which has its source in the upper confines of the district, traverses it in a winding course of upwards of 60 miles, bisecting it longitudinally, and afterwards wafts the trade of Glasgow to the ocean. This tract is otherwise called the county or sheriffdom of Lanark. It is situated between 55° 22' and 55° 58' north latitude, and between 3° 15' and 4° 19' west longitude. It is in the centre of the country, between the Atlantic and German Oceans, and is bounded by the counties of West and Mid Lothians and Peebles on the E., by Dumfriesshire on the S., by the counties of Ayr and Renfrewshire on the W., and by those of Dumbarton and Stirling on the N. The greatest length from N. to S. is

about 47 miles, and the greatest breadth from E. to W.
about 32 miles. The square contents are perhaps nearly
870 miles, equal to 556,800 English acres, or nearly
445,440 of the ordinary Scotch measure.

SECT. II.—DIVISIONS.

THIS district is subdivided into three inferior divisions
called *wards,* each under the particular jurisdiction of a
substitute, appointed by the sheriff depute of the county.
The upper ward, of which the ancient burgh of Lanark
is the chief town, comprehends the parishes of Carluke,
Lanark, Carstairs, Carnwath, Dunsire, Dolphinton,
Walston, Biggar, Liberton, Lamington, Coulter, Crawford, a small part of the parish of Moffat, the rest of
which is in Dumfries-shire, Crawfordjohn, Douglas,
Wiston and Roberton, Syminton, Covington, Petti-
nain, Carmichael, and Lesmahagow. The middle ward,
of which the town of Hamilton is the centre, comprehends the parishes of Hamilton, Blantire, Kilbride,
Avondale, Glasford, Stonehouse, Dalserf, Cambusnethan, Shotts, Dalziel, Bothwel, East Monkland, and
West Monkland. The lower ward, lying immediately
around the city of Glasgow, besides the country or barony parish of Glasgow, contains the parishes of Calder,
Cambuslang, Rutherglen, Carmunnock, Govan, and a
part of Cathcart ; the remainder of this last parish being
in the county of Renfrew.

There was a large map of this county published by Mr.
Ross, in the year 1773, and an abridgement of it has
been since made on a small scale. As no plan of the
county has been taken for this Report, which, for the
sake

sake of the many new lines of roads lately drawn, would have been desirable, the reader must be referred to these maps.

SECT. III.——CLIMATE.

THERE being a great difference in the situation and altitude of this county, not less variety takes place in the state of the weather than in the surface and soil of different parts. Besides that inconstancy of climate, to which the island in general is subject, there are circumstances in the local situation of the county, which occasion considerable variations of the weather, in the different parts of it; so that to be able to give a good account of these, would, perhaps, be to give a tolerable medium state of the weather in Scotland. The lower end of the county is situated in a narrow isthmus, not much more than 30 miles broad, between the Forth and the Clyde, which open gradually to the sea, on each side of the island, admitting the temperate breath of the sea breeze. The wind is computed to blow, about two-thirds of the year, from the S. W. and W.*, over a vast ocean, where no land intervenes, to prevent it from coming to the coast, saturated with the moderate temperature of that element. The winds from the easterly

---

* That this is the most frequent and most forcible wind, is evident from the appearance of the trees, the tops of which generally incline to the N. E. And it is no less remarkable, that they put out the longest and strongest root towards the S. W., in order to support them against the most frequent attacks; so that, when a N. E. storm happens, triple the number of trees are blown down, which fall under as great blasts from the opposite quarter.

B 2   points,

points, which, coming from the Continent, over a narrow sea, are sharper, blow less frequently, and their
force is somewhat broken by the high land on the east
side of the county, so that the cold damps, called *Eas-
terly-Hars*, so prevalent on the east-coast, seldom arrive
here: consequently the cold is moderate. Intense frost
is seldom of long continuance; and deep or long lying
snows are rare: hence there are few spots on which the
verdure of the year is longer protracted. On the other
hand, the most frequent winds coming over so wide an
ocean, are fraught with vapour, which frequently over-
clouds the sky, cools the air, and renders the Summer's
heat less intense; so that it is frequently scarce sufficient to ripen the fruits of the earth. These vapours, intercepted by the neighbouring heights in the counties of
Renfrew and Dumbarton, fall in frequent and heavy
showers, on the northern parts of the county. In going
up the Clyde the surface flattens; scarce a mountain,
between the river and the ocean, raises its head to catch
the clouds; and the current of air passing, without interruption, across the country, the rain is less*. This
circumstance occasions other variations of the weather in
different situations. The clouds, in passing over, often
water the higher grounds on the eastern and western
sides of the county; and, buoyed up by the dense air in
the hollow, leave the middle parts dry. The effects of
these, and other meteorological phenomena to be
now mentioned, on agriculture and vegetation, merit
attention.

---

* As a proof that the height of the land between the Clyde and the
west coast is very moderate, there are places in the middle ward, on
the north side of the river, from 100 to 150 feet above the level of the
sea; from whence all the heights of the isle of Arran, above fifty miles
distant, may be seen in a clear day.

But

But in discussing this subject, it may be proper here to premise, that the motions of the air, in calm weather, which sometimes vary, in the course of a few days, round all the points of the compass, being only breathings from the nearest clouds, which occur during the cessation of the more general currents, are not characteristic of the climate of any particular tract. The same thing may be observed of the rains which accompany thunder; they being neither dependent on the general winds, nor the position of the land, but falling indiscriminately on any place, over which electrical clouds happen to be suspended; though they are no doubt most frequent and copious near high summits which attract these clouds.

The most frequent wind, as has been already said, is from the S. W.; next to it is the N. E., which, for the most part, is accompanied with fair weather. The heaviest and most lasting rain, but not the most frequent, is from the S. E. The wind seldom blows long from the S. without bringing rain; and this rain is heavy, but of short continuance. The rain from the W. and S. W. comes in repeated showers, between short intervals of fair weather; and the greatest quantity of rain, here, comes from the latter, which, as the wind blows much from that quarter in the beginning of the year, generally drenches the ground greatly before seed-time. Rains from the N. W., N., and N. E., are neither frequent nor heavy, but sullen and unnourishing. The N. E. wind is most frequent in the months of April and May: it sometimes sets in in March, and is of great importance, in preparing the ground for the reception of the seed.

In a country, where there is almost every where an under stratum, through which no water can filter, in

spite

spite of every attention to draining, the land is soaked
with water, which can only be drawn off by means of
exhalation; and ground of this kind is not in a condition
to exert its powers, till the cold, sluggish moisture of
the winter is exhaled. When a course of dry weather
does not happen, therefore, in due season, the seed
time must either be deferred to a late period, or the seed,
committed to the crude soil, will make but a slow and
imperfect progress. This is one cause of the lateness of
the Lanarkshire harvests. The air which overspreads
the hollows, being dense and dry, contributes to quicken
exhalation, so that the low lands dry sooner than those
on the heights; upon which a lighter, moister air floats,
and flying showers oftener fall. The seed-time is therefore later, or the preparation less perfect on the heights
than on the plains; which contributes, with other causes,
to render the harvests less early and less mature on the
former, than on the latter. Perhaps, too, there is a
principle mingled with the moist air of the heights, still
more inimical to vegetation, than the moisture which the
surface soil retains. Those huge masses of peat earth,
with which the county abounds, are so cold and torpid
that it is not uncommon, after a hard Winter, to find
the frosty congelation, in large bodies, still remaining
in them, when the Summer is far advanced, and the
heat has been considerable. May we not then suppose,
that the cold moisture arising from them, in exhalations,
before it acquire the temperature of common air, may
cool the earth, and benumb the vegetables over which
it passeth? However this may be, it is certain, that the
high lands, where these bodies of peat earth abound,
are less fertile than they have been in early times; neither the stately oak, nor its accompanying brush-wood,
now appear, where they had once flourished abundantly;
and

and now lie buried together under the torpid chaos: nor can the utmost industry restore them in that vigour, with which in former ages they have spontaneously grown. The oak particularly, now planted on the best soil, in the midst of a thicket, will barely live; while perhaps, within an hundred yards, one of a magnificent size may be found lying on the spot, where, it may evidently be seen, by its roots still fast in the earth, it had been produced.

Something to the same purpose may be observed respecting the annual productions of the earth. There are many places to be found, on the ridges on both sides of this county, which seem once to have been accurately cultivated, and now for a time left neglected. Judging from the appearance of the country where these places are, it is presumable that the climate in which they are situated is not adapted to the growth of corn; and it is probable they have been abandoned for that reason. But it is plain the cultivators had not always found this to be the case; for the appearance of the places in question is not like that of slight attempts, rashly undertaken, and quickly abandoned: but like that of fields, which have been for a considerable time under attentive culture*. How shall we account for this failure of ferti-

---

* Besides the increased cold, occasioned by the increase of peat moss, one may suppose other causes of this desertion. For example, it is said the wild grey oats still continued to be sown on some of the worst land in this county, about the beginning of the eighteenth century; and they are yet sown in some places of Galloway, and the North Highlands. This plant is hardier, and thrives on worse land, than the cultivated oats. The wild oats might have been cultivated on the places alluded to; and when the other oats came into general use, the fields which are fit only for the former, might be neglected. But this will hardly satisfy a person who has carefully examined the country, and discovered many other marks of diminished fertility.

lity?

lity? The climate, upon the whole, is much the same as it was from the earliest notices we have of it. The illustrious biographer of AGRICOLA *, about the time when that General was erecting ramparts upon the northern confines of this county, characterizes the Caledonian weather thus: " Cœlum crebris imbribus ac nebulis fœ-" dum: Asperitas frigorum abest;" which corresponds pretty much with the account above given, of the climate of the lower ward. It must therefore arise from some such local cause as that which has been stated. It is natural to suppose, that from those large beds of spongy matter, a considerable quantity of moisture charged with the chilling cold it has contracted, must be daily evaporated. This must either be dissipated in the surrounding atmosphere, increasing the cold all around the places from which it is drawn; or, creeping unmixed along the surface, produce still more baneful effects, wherever it spreads. In either case, peat mosses, besides being of little use themselves, are highly unfriendly to the general vegetation of the country; and as they are still accumulating, it is melancholy to anticipate, to what height their malignant influence may arrive.

We now return to take notice of other effects of the climate on vegetation. The greatest droughts are in the months of May and June. When the weather is dry in the latter end of May, it commonly continues so in June, till thunder brings rain†. The violent exhalation, in these months, suddenly binds up the wet clay soil, making it so hard that the roots of the plants cannot extend in it; so that vegetation is almost suspended till the rains

* TACITUS.

† From the 8th of May to the 18th of June, 1788, not a drop of rain fell in the middle ward.

come.

dome. The ground is generally well watered towards the latter end of June and beginning of July, and then a great growth commences; and as the latter end of Summer is generally more rainy than the beginning of it, this growth is protracted till the season is far advanced, and occasions a late harvest. The drought in May, which is often accompanied with a cold wind, greatly checks the rising of the pasture, but is still more injurious to the hay crop, which, when stunted at this time, never again recovers.

Proceeding up the Clyde, the island becoming broader, and the situation more distant from the mouth of the river, the influence of the sea air diminishes. The eminences in the upper ward being more frequent, and of greater altitude, give more interruption to the current of air from sea to sea; and the climate is more similar to that of a continent, the Summer's heats, and the Winter's colds, being more steady, and more intense*. There is more rain above the falls, than in the middle ward; but from the nature of the soil, it is seldomer injurious to the husbandman, excepting in the time of harvest; nor are the Summer's droughts so hurtful. Ascending to the upper extremity of the county, another change again takes place. The highest summits intercept the clouds, and occasion frequent rains; frequent mists hover on the hills, obscuring the sky and cooling the air; the Summer heats are frequently interrupted by cold

gusts;

---

* The following Table is extracts from the accurate observations of an ingenious friend, residing in the low grounds of the upper ward, on the extremes of Summer heat and Winter's cold, for a number of years past, contrasted with the greatest heat and cold on the same days; extracted from a very exact journal of the weather kept near the medium height of the middle ward, and will best illustrate what is here stated. The degrees are marked by thermometers on FAHRENHEIT's scale.

TABLE,

gusts; the Winters are cold and tedious, long retaining on the surface the heavy snow which falls in that tract.

Although the general character of the climate is the same now as in former ages, it is certain that courses of good and bad seasons repeatedly occur, which materially affect the produce of the earth. It would be a curious inquiry, were we able to trace these revolutions for a considerable series of years; and might probably lead to some discovery, whether there be any fixed law in Nature, by which they are brought about. But perhaps, there are no records to be found, which could be depended

TABLE,

Showing the extremes of heat and cold in the upper and middle wards' on the days under-mentioned.

HEAT.					COLD.				
Greatest Height of the Thermometer.					Degrees below the Freezing Point.				
Year.	Month.		—	Up.	Mid.	Year.	Month.		—	Up.	Mid.
1785	June	25	—	80°	76°	1783	Dec.	30	—	30°	24°
		26	—	80	77	1784	Jan.	21	—	31	17
		27	—	85	87			23	—	35	20
		28	—	85½	84			25	—	30	10
		29	—	82	80			27	—	23	21
		30	—	82	80		Dec.	13	—	28	16
	July	26	—	80	74	1785	Dec.	29	—	25	11
1788	June	17	—	79	75			30	—	30	15
1791	July	2	—	79	56			31	—	41	24
		3	—	81	66	1786	Jan.	1	—	43	40
1792	Aug.	8	—	78	70			2	—	41	19
1793	July	9	—	78	72		Dec.	20	—	32	18
		10	—	84	71	1788	Dec.	16	—	38	32
		11	—	77	66	1789	Jan.	11	—	27	16

Note—On the 3d of October, 1782, the thermometer stood 16° below the freezing point, and on the 4th, 12°, in the upper ward. The extreme cold of these days is not marked in the middle ward journal. The earth was frozen, but at nine A.M. the thermometer was above the freezing point.

upon,

upon, for this purpose. It is said there are registers of the price of grain, which tend to show that a course of bad seasons happened about the end of the 16th century: and it is well authenticated, that six or seven years, near the end of the 17th, were very unfavourable. By the concurring testimony of many persons living, and lately dead, we are assured, that from near the beginning of the last century, to the year 1740, the seasons were mostly very favourable. From thence to 1756 inclusive, many unfavourable ones occurred. The year 1757, though the spring was backward, was a warm summer, and certainly had a good crop; for the price of barley fell that year from 1l. 6s. 8d. to 10s. 6d. per boll. From this to 1772, there were several pretty favourable seasons; and few have happened since, in which the fruits of the earth have not suffered from unseasonable frosts in spring or autumn, or both. The years 1779 and 1781, were the most favourable, and the crops early and good. The severity of the year 1782, is too memorable to be forgotten; and the frosts in the beginning of summer and in autumn, have been very prevalent ever since. It may be proper here to subjoin the degrees of cold, by which different vegetables are injured.

TABLE,

Of the Degrees of Cold beneath the freezing point, which injure different vegetables.

	Degrees.
Potatoes, either in spring or autumn, the leaves affected and growth checked, by	2
Ditto, ditto, the stem attacked, and its farther growth prevented, and if in autumn, the root makes no more progress, by	4
Green pease and barley, when the ear is just become milky, spoiled by	4

Beans,

Degrees.

Beans, when green or wet, by 5

Oats, when the ear is milky, by 6

Ditto, when green, and the ear watery on porous } 7
  soils, will stand, without being much hurt

Ditto, ditto, on firm clay, ditto 10

Clover, in the tender leaf, hurt by 3

Ditto and rye-grass, the crop in May or June, } 9
  ruined by ....

Turnips on the ground in winter, injured by 24

Ditto, ditto, totally destroyed, by 30

Note—That frost hurts plants soonest, when the air is still and the dew great, or when it comes on immediately after rain; and the injury does not happen unless the frost continues till the sun be above the horizon. When the frost goes off before morning, and the sky is overcast with fresh clouds, plants escape unhurt, though they may have suffered a pretty severe frost during the night.

We shall now conclude this article, by subjoining the following Table, containing an abstract of the average heat, and number of dry and wet days every month of the year, for two periods of five years each, at 20 years distance, which, it is hoped, along with what has been said, will convey a pretty distinct idea of the state of the weather.

TABLE,

TABLE,

Showing the average height of the Thermometer, and number of dry and wet days in the Middle Ward of Clydesdale, for five years, beginning with 1768, and five years beginning with 1788.

	April.			May.			June.			July.			August.			September.			Amount.		
Years.	Ther	Dry	Wet	Ther	Dry	Wet	Ther	Dry	Wet	Ther	Dry	Wet	Ther	Dry	Wet	Ther	Dry	Wet	Ther	Dry	Wet
1768	50	23	7	59	29	2	62	28	2	64	19	12	63	25½	5½	57	19½	10½	355	144	39
1769	49	24½	5½	56	26	5	58	20	10	65	24	7	61	21	10	58	18½	11½	347	134	49
1770	44	25	5	56	25½	5½	58	21½	8½	63	23	8	66	27	4	61	20	10	348	142	41
1771	46	26½	3½	56	22	9	64	27	3	65	24½	6½	61	22½	8½	55	25	5	347	147½	35½
1772	47	27	3	55	26	5	63	24	6	65	23½	7½	62	24½	6½	58	23	7	350	148	35
Sum.	236	126	24	282	128½	26½	305	120½	29½	322	114	41	313	120½	34½	289	106	44	1747	715½	199½
Aver.	47	25	5	56	26	5	61	24	6	64	23	8	62	24	7	58	21	9	349	143	40
1788	41	22	8	54	27	4	50	27	3	55	20	11	62	25	6	54	20	10	316	141	42
1789	43	22	8	52	24	7	55	18½	11½	58	17½	13½	61	27	4	55	22	8	324	131	52
1790	40	24½	5½	50	25	6	54	23½	6½	56	24	7	55	18½	12½	50	20½	9½	305	136	47
1791	44	21	9	47	21½	9½	56	22	8	56	21	10	58	23½	7½	56	27	3	317	136	47
1792	46	23	7	47	22	9	53	22	8	58	21	10	59	25	6	49	20	10	312	133	30
Sum.	214	112½	37½	250	119½	35½	268	113	37	283	103½	51½	295	119	36	264	109½	40½	1574	677	238
Aver.	43	22	8	50	24	7	53	23	7	57	21	10	59	24	7	53	22	8	313	136	47

TABLE CONTINUED.

	January.			February.			March.			October.			November.			December.			Amount.		
Years.	Ther	Dry	Wet	Ther	Dry	Wet	Ther	Dry	Wet	Ther	Dry	Wet	Ther	Dry	Wet	Ther	Dry	Wet	Ther	Dry	Wet
1768	31	23	8	39	16	13	43	27	4	53	23½	7½	40	18½	11½	35	19	12	241	127	56
1769	34	25½	5½	42	22	6	42	23	8	52	24½	6½	42	23	7	43	21½	9½	255	139½	42½
1770	37	24½	6½	42	23½	4½	34	26	5	49	25	6	37	18	12	38	20	11	237	137	45
1771	30	27	4	41	25	3	37	26	5	51	18	13	46	23½	6½	43	25	6	248	144½	37½
1772	31	23	8	27	24	5	39	25	6	54	21½	9½	46	16½	13½	40	26	5	237	136	47
Sum.	163	123	32	191	110½	31½	195	127	28	259	112½	42½	211	99½	50½	199	111½	43½	1218	684	228
Aver.	32	25	6	38	22	6	39	25	6	52	22	9	42	20	10	40	22	9	243	137	45
1788	37	28½	2½	36	25½	3½	36	28½	2½	45	29½	1½	41	24	6	27	28	3	222	164	19
1789	32	24	7	38	20½	7½	34	26	5	46	21	10	38	22½	7½	40	22	9	228	136	46
1790	35	26	5	42	23	5	42	27	4	46	25½	5½	36	24	6	34	23	8	235	148½	33½
1791	35	18	13	36	20½	7½	41	27	4	47	22	9	38	21	9	30	27	4	227	135½	46½
1792	31	26½	4½	39	24	5	38	22	9	42	22	9	41	24	6	34	20	11	225	138½	44½
Sum.	170	123	32	191	113½	28½	191	130½	24½	226	120	35	194	115½	34½	165	120	35	1137	722½	189½
Aver.	34	25	6	38	23	5	38	26	5	45	24	7	39	23	7	33	24	7	227	144	38

This Table is abstracted from the journal of a learned and ingenious gentleman, who has regularly kept it, in the centre of this district, for 27 years; during which time he has accurately noted the state of the weather, the rains, snows, &c. which happened every day, the rise and fall of the mercury in the barometer and thermometer, with other meteorological remarks. In this abstract, the average heat is taken from the height of the thermometer at nine o'clock in the morning.

When rain, hail, or snow, had fallen for the greatest part of the day, it is called a wet day. If there were only some showers in a day, it is made half wet, half dry; but when there were only some slight drizzling, two of such are made half a day. The six months from March to October, the heat of which is necessary to raise the vegetables cultivated by husbandmen, are in the first part of the Table; and the other six months, in the beginning and end of the year, in the second. But though this shows something of the time and continuance of rain, it conveys no idea of the quantity which fell; nor has it been ascertained in this country, except by the learned professor of Natural Philosophy of the University of Glasgow, Mr. ANDERSON, who has invented, perhaps, the most ingenious and accurate rain gage that has yet been known. It receives the rain at a little more than 100 feet above the level of the sea, and has been regularly kept since the year 1781; previous to which, the rain was measured by an old rain gage. This gentleman has been so kind as to communicate the quantity of rain fallen at Glasgow, for each month and year since his apparatus was erected, from which the quantity, for the last five years of the foregoing Table, is extracted. For the reasons already assigned, the rains are no doubt more copious at Glasgow, than in the flat country some miles up the river;

and

and do not always fall at the same period on both places: yet it may be observed, that in the months which have the greatest number of wet days, in the one, the quantity of rain in the other is generally greatest. The depth of water, which had fallen in each month, is marked in the columns under the name of the month, in inches and decimal parts.

TABLE,

Of the Rain which fell at Glasgow, for the Five Years under-mentioned.

	Jan.	Feb.	March	April.	May.	June.	July.	August	Sept.	Oct.	Nov.	Dec.	Sum.
	In. Parts	In. Parts	In. Parts	In. Parts	In. Parts	In. Parts	In. Parts	In. Parts	In. Parts	In. Parts	In. Parts	In. Parts	In. Parts
1788	2.4	.98	1.72	1.68	.916	1.251	2.971	1.29	2.8	1.7	1.57	.15	19.428
1789	3.9	3.91	.86	1.686	1.85	1.82	4.68	.67	3.875	2.78	2.67	6.5	35.201
1790	2.07	2.81	.51	2.03	2.02	1.06	3.3	3.774	4.373	3.23	2.	4.29	31.467
1791	3.97	2.4	1.12	2.219	2.783	2.3	2.773	2.835	3.315	4.082	2.272	6.624	32.105
1792	1.71	1.721	3.21	2.32	Six months here in cumulo 15.99.						4.	3.106	35.945
Sum.	14.05	11.821	7.42	9.915	7.569	6.431	13.724	8.569	14.363	11.792	12.512	20.67	154.146
Aver.	2.81	2.364	1.584	1.983	1.892	1.607	3.431	2.142	3.59	2.698	2.502	4.134	30.829

SECT. IV.—SOIL AND SURFACE.

THIS county is so extensive, and the surface so greatly diversified, that it would be impossible to give any tolerable map of soils. The upper ward, which is nearly two-thirds of the whole county, is mostly mountainous, or at least hilly and moorish; and from the nature of the soil, and the great elevation of the country, not capable of much agricultural improvement : between two-thirds and three-fourths of its extent may be comprehended under this description. The wide parishes of Crawford and Crawfordjohn, the greatest part of Lamington and Coulter, are high and rugged. Three-fourths of Douglas and Lesmahagow, on the one side, and of Dunsire on the other, are either moorish heathy land, or covered with beds of peat earth, yielding but little useful herbage. A considerable part of the parishes of Carluke, Carstairs, Lanark, Carnwath, Walston, Dolphinton, and Biggar, is of much the same quality.

At the head of the county, where it bounds with that of Dumfries, the country is very mountainous. The great edifices of Nature are so closely huddled together, that their grandeur is lost to the eye of the beholder. When he traverses the hollows, only the sides of the nearest mountains are presented to his view ; if he climbs an eminence, he sees nothing but a confused group of rugged tops, with the naked rock frequently appearing among the herbage. The elevation of this tract is very great, the site of the village of Leadhills being computed to be 2000 feet above the level of the sea, and the top of one of the *Lowthers*, a ridge of hills near that place, has been found to be 1100 feet higher, making the height of the summit 3100 feet.

In

In proceeding down the Clyde, the prospect opens; the hills stand at a greater distance from one another, and the ascents are less abrupt : villages, farms, corn fields, and plantations of trees, appear among the eminences; and the mixture of hill and dale forms a scene at once simple and sublime. The nature of the soil, however, is not always more fertile, as the elevation becomes less. The pastures on the heights of Crawford are superior to those of Douglas or Dunsire. The former are upon hard rock, and many of them pretty dry, covered with a thick mixture of short heath and sweet grasses; sometimes a close verdant carpet, with very little heath. The latter are frequently wet and spongy, and their herbage thin and coarse. As the hills decline in height, the rock more seldom appears on the surface; and beds of gravel, of a considerable depth, are sometimes seen.

Tintoe, or Tintock, is the last great hill to the north; and forms the boundary of the hilly district on that side. The height of this hill is about 2260 feet above the level of the sea; and the medium height of the arable land around its foot from 600 to 700 feet. From Tintoe, the face of the country is softened down to gentle elevations and depressions; and the Clyde slowly glides, with many windings, through a tract of beautiful meadows, for above a dozen of miles, till it arrives at the head of the celebrated falls. From thence, it rushes from cataract to cataract, foaming among the fragments of rock for about six miles; and regains its quiet bed and gentle motion in the lower part of the upper ward. The uppermost fall is that of Bonnington, where the river precipitates over the edge of a perpendicular rock. The height of this, including a little one immediately above it, is about 30 feet. The second fall is that of Corhouse, over which the river dashes from one ledge of a shelving rock to

c 2   another.

another. Its perpendicular height is 70 feet. Dundaff
fall is 10 feet; and there are three distinct falls at Stone-
byres, succeeding one another, which measure together
about 70 feet.

The principal part of the arable land above the falls, is
in the parishes around Tintoe, which lie along the side of
the river. The soil of the meadows by the river side,
formed by the slime deposited in floods, is of the nature
of carse ground, with a greater or less mixture of sand,
in proportion to the quicker or slower motion of the
stream, by which it has been deposited. These meadows
are very fertile, and are still receiving additions from
the inundations of the river. But this cause of fertility
is also the cause of frequent and considerable damage,
the inundations sometimes destroying the crops, shifting
the course of the river, carrying away the rich soil, and
leaving beds of sand and gravel in its stead. The uplands,
with the exception of some few places where springs
arise, are dry and very fertile, generally of a light and
friable quality, with an under stratum of sand or gravel
of a considerable depth. Some places occur which have
not been sweetened by culture, and have a sterile appearance:—others, where the soil is either moorish and
spongy, or somewhat of an argillaceous quality, having an
impervious under-stratum. This last is more frequently
the character of the land which lies distant from the
river.

From the beginning of the falls downward, the style of
the country is altogether changed. Instead of the basaltic
or whin rock standing in perpendicular columns, the free-
stone rock, lying in horizontal strata, begins to take place.
The subsidence of the land does not keep pace with the
fall of the Clyde. From a mild and calm river, softly
flowing through level meadows and wide expansive banks,

it

it becomes an impetuous torrent, deep ingulfed in a double range of steep hills, seeming impatient of its straitened course. The brooks which fall into it are somewhat of the same character. The rushing waters, the lofty and diversified rocks, the towering summits, the overhanging woods, exhibit altogether a scenery, in which the beauties of Nature are happily blended with her grand and sublime works, and form a proper subject to warm the imagination, and exercise the genius, of the poet and the landscape painter; but it is the business of agriculture to dwell on plainer scenes.

The greatest part of the arable land, in the parishes of Lanark and Lesmahagow, is dry, light, and friable, though much less fertile, somewhat resembling that in the neighbourhood of Tintoe, already described. In the lower part of the latter, the clay soil takes place; and much of the soil of Carluke parish is either of a clay nature, or has a dense argillaceous bottom. A great deal of it is damp, cold, and barren; but some of it is of a very good quality; and that verge of the parish which lies along the Clyde, is not less fertile in soil than rich in picturesque beauties.

Towards the lower part of the upper ward, though the soil in general is less fertile, the country becomes more interesting. Handsome seats, surrounded with well dressed fields, sheltered with clumps and belts of trees, are frequent; villages filled with industrious inhabitants, arise on all sides; and the efforts of those people, for their own accommodation, are continually giving beauty and fertility to some new spots. The course through which the Clyde flows, gradually opens; the river expands, and gently purls over its pebbled bed, through alternate tracts of sloping banks and fertile valleys, ornamented

c 3

sometimes

sometimes with a mixture of orchards and copse-wood, sometimes with tufts of forest trees.

At the commencement of the middle ward, the loftiness of the land is considerably diminished, and it still continues to fall to the N. W. The whole face of the country, when beheld from any distant height, appears like one great plain. Plain, however, is not the character of the county of Lanark. The surface is every where diversified by wavy inequalities, scarce a plain of any considerable extent intervening, except the valleys along the sides of the river; from which the surface, as it recedes, rises irregularly to the highest ridge, on each side, near the confines of the county. The height of these ridges, if a few particular summits of no great importance be excepted, is not more than 700 feet above the level of the sea. The site of the town of Hamilton, on the low ground, in the centre of the middle ward, is computed to be from 100 to 140 feet. The medium height of the cultivated land, will probably be from 250 to 300 feet.

Although there is a great diversity of soil in the middle ward, it is most generally of a clayey nature, with a greater or less intermixture of sand, and very different in colour, conformation, and degrees of fertility*. The bottom is a solid argillaceous substance, sometimes seemingly homogeneous, and lying in regular horizontal laminæ, but oftener of a mixed nature, without the appearance of divisions in any direction, and commonly

---

* Iron, in different combinations, is most commonly the colouring substance in all the soils of the county. And it is favourable or adverse to fertility, according to the nature of the combination. That which approaches next to the metallic state, is the most favourable.

mixed

mixed with little roundish stones of different sizes and appearances. Sometimes a little tract of sandy or gravelly soil occurs; and when a bed of this open quality is of a tolerable depth, the land is dry; but wherever the clayey under-stratum approaches near the surface, which frequently happens, the soil is soft and wet. At some distance from the river is frequently found, lying upon a clay bed, a thin loose soil, much disposed to heave with the vicissitudes of the weather, and very unfit to furnish either nourishment, or a sufficient hold to the roots of plants. The black or grey soil, on the high moorish grounds, is somewhat similar to this, but generally bears a good deal more grass. The water-formed soils in the valleys, by the sides of the river and some of the other considerable streams, differ greatly from all the above mentioned, being naturally more fertile, deeper, and generally less apt to be injured by rain, as they lie upon a bed of open gravel. They differ also from one another, according to the nature and proportions of the materials of which they have been originally composed. The rich mellow earth, which, by some people, has been distinguished as a species of soil, under the name of *loam**, being the residuum of decomposed vegetables continually accumulated, may be a part of any soil, where circumstances have occasioned this accumulation, and will be more or less fertile in proportion to the nature and quantity of the vegetables which have produced it, to the quality of the soil upon which it has been engrafted—to the bottom on which it lies,—and to the exposure and

---

* Loam is nothing else than sand and clay blended in different proportions, containing also a considerable share of the least soluble parts of the spoils of decayed vegetables, according to the nomenclature of modern chemistry, called oxide of carbon.

elevation

elevation in the atmosphere. Accordingly, we find the loamy soils, in different places, possessing very differe: degrees of fertility. They are either in those fields ne: farm-houses, where the farm dung has been long r: peatedly applied, called *old crofts*, or on dry bottom: where leaves of trees and sweet herbage have long bee: allowed to rot, or where the same substances have bee: carried down by streams, and lodged in the valleys. Th: best of these are where the original soil is at the san: time firm and friable, and the bottom open and dry.

Besides the above-mentioned kinds of soil, a conside: able part of the surface is covered with beds of pe: earth, which have overwhelmed the original soil, an: assumed its place. These are chiefly of two kinds, th: one generally of a moderate thickness, composed sole: of decayed vegetables, nourished by a cold watery soi: and damp atmosphere. As those vegetables, which ai: nourished by a genial heat and kindly soil, are quickl: susceptible of the putrid fermentation, and in rotting fa: into loam; so those that thrive in cold moisture, hav: something in their nature, which, in a great measure pre: serves their form and bulk, even in decay; and by th: growth of one year above another, through the lapse c: time, in a neglected country, they accumulate into bec: of this inflammable earth. Almost all the tribe of *Mosse.*: (Musci) and some other coarse aquatic plants, are of thi: nature; and hence, perhaps, those accumulations hav: got the name of *moss*. When those beds lie on lov: ground, where water can carry particles of heavy eart: upon them, they are thus rendered less porous, and pro: duce a considerable quantity of grass; when they are oi: rising ground, heath (*erica*) and deer-hair (*scirpus cespitosus*: are the chief productions. Some of these have beei: drained, pared and burnt, and produced pretty good crop:

oi

of corn and grass. The other kind lies generally in plains or hollows among the eminences, and is generally of a great depth. It is evident that all of them, in former times, have been forests of large trees*; some of which having fallen down across the water-course, and intercepted whatever was brought along by the stream, must have formed dams; which stopping the water, would convert the whole into a standing pool. In this manner the trees would die, as the standing water prevailed over the roots, and the whole forest at length became prostrate.

Over these, mosses, and a variety of aquatic herbage, have grown, from age to age, till they are swoln to great masses of spongy matter; these are called *Flows*, or *Flow-mosses* †; they are much more in this district than in the other, and, in a country where coals abound, of very little use, neither producing much esculent herbage, nor admitting cattle to go safely in quest of it. At the same time they may almost be said to be irreclaimable. Only two methods have hitherto been proposed for this purpose; the one is to lead some brook to the place, and to float away the spongy surface in the water, by cutting it into fragments with spades. The bottom, after being properly drained and cultivated, is capable of producing

---

* Some of them appear to have been lakes fringed about with a margin of coppice, as a body of water of several feet deep is frequently found under the peat towards the centre of those grounds. This is called by the country people sole water.

† Some attribute the increase of these spongy mosses to the continued growth of a species of muscus, called *sphagnum palustre*. It is certain, that the sphagnum, which rises in large tufts, frequently contribute; but many other of the musci and algæ make part in the formation of peat, besides some larger aquatic herbage. And acres of peat, still increasing, may be found, on which a single tuft of sphagnum does not appear.

corn

corn or grass. The other is to drain the whole sufficiently,
to smooth the surface by paring, burning, &c. and then
to give a thick top-dressing of any kind of heavy earth,
such as sand, small gravel, or clay, and perhaps clay is
the best. This soon covers the whole with verdure, and
may be repeated from time to time, as occasion requires.
It is an improvement, to stir the surface by turning it with
the plough, after it has acquired solidity, so as to mix
and incorporate the different kinds of earth. The first
method has not been practised in this country, except in
the making of drains. The second has been followed in
some few instances, with considerable success; and there
is the more encouragement to continue it, that when the
most barren clayey substance, dug at a great depth below the surface, in which all vegetables die while it lies
in a mass, when spread over the mossy substance, will,
in a year or two, produce white clover and other sweet
herbage *. But, alas ! what labour would be sufficient
to subdue the sterility of such extensive wastes ? the
mosses of the middle ward being, as nearly as can be computed, about 42,000 acres, which is almost a third of the
whole extent of the district. Spots occur on the lower
ground, on which another kind of mossy earth is found,
formed by cold springs, arising on the surface, and encouraging the growth of those plants, of which this kind
of earth is composed ; but these are of very inconsiderable extent.

　Having now mentioned the different soils to be found
in the subdivision of the county of which we are now
treating, it may be proper, in order to give a better idea

---

* An important discovery has been lately made, in a neighbouring
county, with respect to the reclaiming of mosses ;—some notice of
which shall be taken in a future chapter.

of

of it, to take a more particular view. The highest ridge
on the north side runs along the eastern extremity of
Cambusnethan parish, through the middle of that of
Shotts, where this ridge is pretty high and rocky, and
thence through East Monkland parish, declining a little
as it proceeds westward. In these three parishes, particularly in that of Shotts, lies the greatest part of the *mosses*
to be found on this side the river. Much of the rest of
the soil along this ridge is moorish, coarse, and wet. All
of it, however, is not of so bad a quality. Along the
Calder, which divides the parishes of Cambusnethan and
Shotts, there is a tract of pretty good soil on both sides.
Near the head of this stream, it is light, sandy, or gravelly, and pretty dry; farther down it becomes a strong
clay. Many fields in East Monkland, though high, are
tolerably fertile; some inclined to sand, some to clay, a
considerable part is of a grey moorish soil, and some of
a mixed nature. This tract is interspersed with spots of
pasture and marshy meadow, saved annually for hay.
The opposite ridge, beginning on the parish of Avondale,
is a continuation of the hilly range, which divides the
parishes of Douglas and Lesmahagow from Ayrshire, and
runs from Avondale, through Kilbride, Cambuslang,
and Carmunnock, to the county of Renfrew. The
wilds here are much more extensive than those on the
north side of the Clyde. Through a tract of twelve
miles long, and sometimes a considerable breadth, there
is scarce any thing but mosses, and benty, or heathy pasture, very wet and coarse; a mixture here and there is
somewhat drier, and may be called green. The arable
land of Avondale is but a small proportion of the whole,
lying in the lower part of the parish. It is mostly of a
gravelly nature, and frequently encumbered with springs
arising on the surface. Many fields may be said to be
fertile,

fertile, and particularly produce abundance of good grass, when left untilled. Above three-fourths of the parish of Kilbride are arable; the soils are various. On the S. E. boundary is a free soil, lying upon an open hard rock, pretty fertile; more to the N. and W. it becomes a stiff moist clay; on the west side the soil is a happy mixture, and very productive. Through these two parishes, considerable tracts of level meadows are interspersed among the cultivated fields; on some of them the spret *(juncus articulatus)* prevails; others, particularly those which have had the soil enriched by the overflowings of some neighbouring brook, producing abundance of sweet hay. The parishes of Stonehouse and Glasford, along the banks of the Avon, are mostly arable. The banks of this river, from its head, till it has passed those parishes, is destitute of copse wood; and the country has a plain, but not unpleasant pastoral appearance. The soil is tolerably good. In Stonehouse is a good deal of sandy soil, pretty dry, and of a pleasant improvable quality. In Glasford it is more frequently gravelly, mostly dry, but springs sometimes occurring. As the land recedes from the Avon, on both sides, the argillaceous bottom approaches nearer the surface, and is covered, either with a grey moorish soil, or a soft earthy clay, frequently thin and moist. In the high part of Glasford, there is a considerable extent of moss. In all hollows among the high grounds, on both sides of the country, the common rush *(juncus conglomeratus)* prevails very much.

What part of the middle ward remains yet undescribed, viz. the lower part of Cambusnethan, the parishes of Dalziel, Bothwell, and West Monkland on the north side of the Clyde, and those of Dalserf, Hamilton, and Blantire, on the south side, lie pleasantly sloping towards the river on both sides. The length, along the

the banks, is upwards of twelve miles, and probably
near half that in breadth, and is perhaps a tract not inferior in beauty to any other in Britain.   At the head of
this tract the banks of the Clyde have already expanded,
and continue to open downward to its confluence with
the south Calder, admitting valleys of varied breadth
along the sides of the river.   Here again the scene varies,
and bold banks, in the parishes of Bothwell and Blantire,
hem it in on each side.   From thence they expand and
contract alternately to the extremity of the county, ex-
hibiting every where a beautiful variety.   The same great
materials,—flowing waters, winding valleys, and swell-
ing banks, form the ground-work of the landscape, both
above the falls and in the lower part of the county; but
the finishing of the one is entirely different from that of
the other.   In the former, nature appears in the elegant
simplicity of a handsome undress; in the latter, magnificently arrayed in her richest ornaments.   The soil and
climate seem to be much disposed to the growth of
wood, and spontaneous copse woods every where fringe
the hanging banks.   Besides the estates of great landholders, much of the land is parcelled out in moderate
and small properties.   The industry and judgment which
so many people of all ranks have exerted, to shelter their
properties, and adorn their places of residence, have dispersed, over the face of the country, groups of trees,
appearing in a beautiful disorder, as if scattered by the
hand of chance.   Numerous villages and hamlets contribute to enrich the scenery.   The labours of a number
of husbandmen, employed in the improvement of the
fields, has produced a verdure which smiles almost perpetually in different corners, to whatever quarter the eye
is turned.   Orchards embosomed in woods, stand all
along the Clyde, by the foot of the rising slopes:—thus
that

that beautiful variety, which the face of the country
has received from the hand of Nature, is every where
heightened and improved*.

The different kinds of soil found in this tract, have been
already enumerated. The valleys are very fertile, but bear
a small proportion to the whole. Clay is the most prevalent soil, and a great deal of it is very productive, with attentive culture and favourable seasons. Sandy and gravelly soils are rare. The loose heaving soil above-mentioned, is to be found in some of the higher grounds, and
is of the worst quality, perhaps, of any. There are no
mosses, except on that side of the West Monkland parish
which lies farthest from the Clyde. In this parish there
is more sandy soil than in any other part of the tract.
But what proportion there may be of each of these kinds
of soil, in a country so diversified, it is scarcely possible
to form a judgment.

The under ward is a very limited district, but having
the city of Glasgow situated in it, a very important one.
The banks of the Clyde, though abounding less in natural
beauties than those above, are still more highly ornamented, being planted all along with handsome villas,
the summer retreats of the wealthy inhabitants of Glasgow, who, in their relaxations from business, have highly
improved a considerable part of this tract. A rocky eminence, called *Dichmount*, occupies a part of the parish of
Cambuslang. The soil upon and around it is light and
stony; that of the rest of the parish is mostly clay, excepting that border which lies along the Clyde. The
soil of Carmunnock parish is much the same with that
of Cambuslang, but less improved by culture; and as
the

---

* Beauties of a cast somewhat more romantic, on the banks of the
inferior rivers of Avon, S. N. and W. Calders, are frequent.

the first does not approach the Clyde, the rich soil along
the river is wanting. But the same ridge of hard rock
runs through it, and is, in some places, almost destitute
of soil. The higher part of Rutherglen parish is clay;
the lower is either sandy, or rich valley ground, along
the side of the river. Except a little clay on a rising
ground, the parish of Govan is sand, originally of a
very poor quality, but now highly improved, and mostly
by the present race of inhabitants, whose local situation
furnishes them, not only easy access to the dung of Glasgow, but strong motives to apply it.

It was here that the late Mr. CROSS, sheriff-depute of
the county, about 40 (before 1794) years ago, made his
experiments in TULL's system of horse-hoeing husbandry,
and cropping continually without manure. This he pursued with great perseverance, and at length was convinced,
that the land required to be recruited from time to time.
He was a man of great judgment and attention, and is
allowed to have been the first who roused that spirit of
improvement in the neighbourhood, which has since been
so successfully exercised. The barony of Glasgow is
wonderfully diversified; the haughs (valleys) of Dalmar-
nock, &c. are fertile to a proverb. The north side of the
parish rises in knolls, the tops of which are frequently
hard and stiff, the bottoms wet and spongy. In many
places the soil originally has been but indifferent, but its
faults are much corrected by an uncommon degree of culture, to which the local situation gives great encouragement. The middle of Calder parish is barren, moist, and
moorish; around the outside is a great deal of good soil,
mostly light, and pretty dry.

Having now traversed the surface of the county, this
article may be concluded, with a few general observations.
Land in the same parallel, other circumstances being
nearly

nearly similar, is always more valuable, in proportion to
the comparative lowness of its situation; thus, for example, land on the elevated fields of Avondale and Kilbride on the one side, and of Cambusnethan and Shotts
on the other, are proportionally less valuable than such
as lie in the low tract between them, insomuch as they are
higher situated; the quality of the herbage being less
succulent and nourishing, and the reproduction slower,
when in grass, and the grain less plump and perfectly
ripened, and the harvest later, when in corn. The same
thing holds good in every other parallel through the
county. The nature of the bottom or under-stratum, has
the same kind of influence as the elevation; moist sweating bottoms producing grain of inferior quality, and
ripening late. These observations, seemingly so trite and
obvious, might have been avoided, had it not been for
what follows. The arable land along the Clyde, above
the falls, seems to be superior to any in the lower part
of the county; not only to those fields, nearly on the
same level, on the ridges of the country, but exceeding,
in real intrinsic fertility, the fine low grounds which are
400 or 500 feet less elevated. The meadows or valleys of
the former, by the river side, are cropped and left in
grass, for a few years alternately, and without receiving
any manure, continue to yield abundant harvests. The
uplands, when properly freed of weeds, are very productive, with half the manure which is found necessary
in the lower part of the county; and the harvests are
generally earlier. One circumstance, however, tends
greatly to diminish the difference of the comparative value of land in these different districts. The spring, but
more especially the autumnal frosts, are more frequent,
and more intense, in the upper country than in the
lower. Those calamitous mildews, sometimes in the
month

month of August, fall down from the sides of the mountains, condense at the bottom, and sweep slowly along the valleys of the upper ward, blasting the harvest whereever they come; while the opener country below, perhaps escapes, and the corn ripens slowly to a tolerable harvest. Such frosts are said to be more frequent these last 30 years than formerly, and particularly since 1782. The narrower the valleys, their effects are more severely felt; so that among the thick clustered hills, near the upper extremity of the county, tillage is almost abandoned.

————

SECT. V.—MINERALS.

As the bowels of the earth are sometimes not less productive than its soil, and contribute considerably to the value of the land; and as the working of mines has important effects on the culture of the surface, some account of the mineral strata seems to be entitled to a place in this Report.

Passing the earthy substances, which lie immediately under the soil, the first thing which attracts attention is the rock. This is of three general classes, namely, the *Sand Stone*, or *Free Stone*; *Lime*, and the *Hard Rock*, known in this country by the name of *Whin*. Each of these is of several different qualities and appearances, and may be ranked under various subdivisions; but it will not be necessary for the present purpose, to take up so much room in this Report, as the mentioning of all these subdivisions would require.

From the lower extremity of the county upwards, to above the falls of the river, some kind of free stone is the most general rock; nevertheless, different ridges of whin

run along through the heights, on both sides, appearing
sometimes on the surface, by which these ridges may be
traced from the rocky mountains downward, through the
whole extent of the county. The free stone is continued
probably through all the plainer parts of the country, but
the regularity of the strata is frequently interrupted, and
one edge sunk deep, while the other is raised. It is found
all along the river, and the streams which fall into it, frequently approaching near the surface, and is much used
in building.

Lime lies in the same tract of country as the free
stone, but is only found, near the surface, in places
which are somewhat elevated, after the free stone, and
many of the strata below it, to be after mentioned, have
skirted out at the surface, and are no longer found. It
is most frequent on the south side of the river, viz. in
the parishes of Kilbride, Avondale, Glasford, Stonehouse, Lesmahagow, Douglas, and the higher part of
Hamilton; on the north side it is found in Carnwath
and Carluke parishes. Both those kinds of rock lie in a
position nearly horizontal.

The great body of whin rock is in the upper part of
the county, standing in perpendicular columns or thin
laminæ on edge. It is mostly so in the lower ridges;
but there are instances of it lying horizontally, like the
free stone and lime. It is of a close texture, and composed of very minute particles. Whether it be, as some
have supposed, the lava of ancient volcanos, or whatever have been its origin, it differs widely in its nature
from free stone; and this difference may probably be
the cause of the difference in the fertility of the soil
between the upper and lower parts of the county, above
noted. The most solid bodies, after they cease to increase, tend, less or more, towards dissolution. Even
those

those hard rocks exfoliate; and wherever the decomposed matter lodges, its fertility is shown, by the deep
verdure which arises; whereas reduced free stone shows
no symptoms of fertility*. It is reasonable therefore to
expect, that the soil, with which different causes must
have contributed to mix a great deal of the former,
should be more fertile than that which has always a considerable mixture of the latter.

Under the free stone lies the coal, for which Clydesdale is celebrated. A number of thin strata, or seams,
as they are generally called, of this valuable fossil, lie
above that which is generally called, around the city of
Glasgow, the *upper coal*; because it is the first that is
found worth digging, to any extent. This stratum is
composed entirely of what is called *rough coal* in Scotland, except a small part near the middle of it, of the
kind called *splint*. 2*dly*, About 16 or 17 fathoms under
this, lies the *ell coal*, so called, because it was first found
of this thickness, but it is frequently from 4 to 6 feet
thick. It is composed of two kinds, called *yolk* and *cherry* coal, with sometimes a parting of splint, and sometimes not. This is a fine caking coal, or what is called
in England, a close burning coal, and is much esteemed
for the blacksmith's forge. 3*dly*, At from 10 to 17
fathoms below the last, lies the seam, called the *main
coal*, so called from its possessing all the good qualities
found in any of the other strata in the county. It contains *rough* coal, *splint*, and *parrot*, or *jet coal*, and is preferred, by the consumers, to all the others, as the most

---

* Calcareous earth is always an ingredient in the hard or whin rock,
and clay, which by some means has lost its cohesive quality, and does not
form into brick, so that the decomposition is well adapted to form a fertile soil.

profitable.

profitable. Its thickness is from $2\frac{1}{2}$ to 9 feet. Sometimes a thin bed of stone is found about the middle of the seam, and the whole thickness is 10 feet. *4thly,* About 13 or 14 fathoms lower, lies the *humph coal.* It consists of *yolk* and *rough* coal, with a thin parting of *splint.* In some places it is without the splint, and unworkable, being much interlaced with thin laminæ of stone, and a kind of petrified black clay, called *blaise.* *5thly,* Below the humph coal lies the *hard coal,* sometimes at 14 fathoms distant. It consists solely of splint and parrot coal, and is found to be the best in the county for the smelting of iron. It is also very good for family use. *6thly,* At a fathom and one half lower, is found the *soft coal,* from 30 inches to 6 feet thick. It is composed of the rough, yolk, and cherry coals, cakes much in burning, and is esteemed a good coal for the blacksmith's forge. *7thly,* About 13 or 14 fathoms below this, lies a coal, called, about Glasgow, the *sour-milk coal.* As it burns slowly, and affords but a weak heat, it is what the miners call a lean coal, and has therefore been but little wrought. There are a number of thin seams of coal under the sour-milk coal, all of a lean quality, and generally much interlaced with laminæ of stone, blaise, or shiver. Under the last mentioned have been found several strata of excellent lime; and more of the thin seams of coal again have been discovered under the lime; but all of them, which have yet been tried, are of a lean quality. The lime found near the surface, on the elevated grounds, is supposed to be a continuation of some one or other of the last mentioned strata found under the coal, which, in the course of their natural rise, have come within reach, in the places where the superincumbent strata of coal, and all its accompanying fossils, did not exist; as lime, worth the working, has never yet been discovered.

discovered above those coal strata, nor in any place till after all the valuable known seams of coal had skirted out at the surface: and any coal, which has been found under the surface lime, is of the same lean quality with that which lies under the deep buried strata of lime.

The above is the number and order of the coal strata, every where along the Clyde, where they are entire. However, this is not always the case. All the mineral strata lie inclining towards the river on both sides, generally somewhat obliquely, and with various degrees and directions of declivity, rising, as they recede from it, till they skirt, or, as it is expressed by miners, crop out, one after another; so that the first coal which is found in some places, is perhaps the third or fourth in the above-mentioned order. These are distinguished by the name of the Clyde strata, or seams of coal, and not only lie along the sides of that river, through all the plain country, but branch out less or more along the principal streams, on some of them to a great extent; lying in the thinly inhabited parts, almost untouched, and affording the public the prospect of an almost inexhaustible fund of fuel, whenever the projected canal shall take place. Besides these, there are other seams of coal in the county, of a somewhat different nature. In the parish of Shotts, a fine yolk coal is wrought, resembling the coal found upon the sides of the Forth, and supposed to be a continuation of one of the same strata. Upon the sides of the Douglas river are extensive colleries, which supply some of the more southerly provinces, where this fuel is wanting. The coal here is also similar to that on the Forth. On the S. W. boundary of the county is coal of the same quality with that wrought

on

on the coast of Ayrshire. It crops out at the surface about the middle of Avondale parish.

There are still some other variations in the coal strata, which merit attention. Near the northern boundary of the county, a species is found, distinguished by the name of the *blind coal*, from its burning with intense heat without flame. This must no doubt have been deprived of its fixed air, by means of subterraneous fire. It is used for the same purposes as coke, and even preferred to coke artificially made, its effluvia being still less offensive. The blind coal is always found under a covering of horizontal whin; and when the same seam is traced, till it comes under the free stone rock, its qualities are entirely changed, and it becomes, in every respect, the common pit coal*. Another species of coal, the qualities of which are directly opposite to those of the last, is found in different parts of the county; it is here called the *candle coal*, or *light coal*, and is said to be the parrot or jet coal of the third seam, in the above enumeration, divested of the other kinds which accompany it, when the seam is complete. But when this is found alone, it seems to be still more exquisitely inflammable; it takes flame the moment it is brought in contact with the fire, and a small fragment of it may be carried about in the hand, like a flambeau, and continued for a long time to give a vivid light.

Iron is another mineral which abounds in this county. It is got only in a petrified state, what is called the ore, not having yet been discovered in such quantities as to attract the attention of the miner. The iron stone is either found in beds of unconnected balls, or in a conti-

---

* The blind coal found in Ayrshire, in the neighbourhood of Kilmarnock, is said not to be covered with whin, but schistus.

nuous

nuous rocky stratum. The balls are the richest. Iron stone is found in the same tract of country as the coal, and is the constant concomitant of that fossil, many beds lying between the different seams of coal; and, it is said, the best lies over the fifth seam, called the hard coal.

At the time the View of this County was last published, iron stone was mined, and the metal brought to its first manufactured state, called pig iron, at three different places of the county. There are now eight blast furnaces in Lanarkshire, employed in making iron. A number of founderies are also occupied in re-casting the metal into a vast variety of utensils. At one of which, cannon mortars, balls, carcasses, &c. are made. And there are now proposals for casting bridges of the same metal. At one forge a considerable quantity of malleable iron is manufactured.

Among the mountains, near the southern extremity of the county, are the well-known lead mines belonging to the Right Hon. the Earl of HOPETOUN. In the same neighbourhood, a vein of copper ore was found, and some attempts made to work it, but without success. Here also a vein of antimony has been lately discovered; how it may turn out, is not yet known. There are abundant quarries of excellent slate among these mountains; but the distance from the populous parts of the country is so great, that there is no encouragement to work them to any considerable extent.

Lime is used for the purposes of building, as a flux in the smelting of iron, and chiefly by husbandmen with an intention to ameliorate the soil. Whatever uncommon increase of fertility the application of it may have occasioned, in different districts, it must only be held as a simple ingredient in the soil, not as administering di-

rect

rect food to vegetables. Quick-lime dissolves in about 1500 times its bulk of water, and in that state kills vegetables. When it becomes effete it is quite insoluble, and cannot pass into the organs of plants. Nor has it any peculiar quality of dissolving the food of vegetables. Wet straw rots faster in garden mould than in effete lime. Its effects on vegetation therefore must be attributed to its nature and properties, and those of the other earths, opposite to each other, acting reciprocally upon one another, and upon the water which falls on them, by which the vegetable food locked up in the soil is set at liberty. It is however a valuable ingredient in soils, as it greatly promotes the growth of the trefoils and other esculent herbage. It has not been well ascertained, what proportion it should bear to the other earths. BERGMAN found a fertile soil contained $\frac{1}{5}$th of effete lime, GIOBERT $\frac{11}{100}$ths, and TELLET $\frac{2}{3}$ths. The Writer of this has found, that more than $\frac{1}{8}$th rather injured than promoted vegetation. Though there is abundance of lime stone in the moorish grounds of the parishes of Lesmahagow, Douglas, &c., the beds which have been found in the cultivated parts of the country are greatly exhausted; and seem every where to be dipping so much under the surface, that it is to be feared the expense of raising it will soon put the price beyond the ability of the husbandman to apply it to advantage. About 260 labourers were employed in the different lime quarries when this Report was last published, and the value of the lime was about 12,500*l*. There is little more lime wrought now than formerly, but the additional expense of raising it makes the price now 14,000*l*.

The numerous buildings which have been carried on for some years past, has greatly increased the demand, and raised the price of free stone: but a just estimate of

its

its amount cannot be made. It must be very considerable, as the price of every block fit for astler is 10*d.*, and every cart load of ruble stone is 1*s.* near Glasgow. For the former period, about 2000 people were employed about the coal mines, and the quantity of coals raised annually was 765,000 tons. The number of labourers at the coal-works are now about 2800, and the quantity of coals raised 1,250,000 tons. The iron works, at the former period, employed about 500 hands, and the annual produce of pig iron was about 3600 tons, by which 36,000 tons of coals were consumed. The blast furnaces and founderies of the county now employ about 1600 people, the produce of pig iron annually is from 9 to 10,000 tons, and the coals consumed about 130,000 tons. The produce of the lead mines varies as formerly, according to the success of the discovery, and may be computed at 34 or 36,000 bars annually; the inhabitants of Leadhills at 1200. The coals consumed at the lead works are brought from Sanquhar, in the county of Dumfries.

Thus the state of the produce of mines, at the different periods, will stand as follows :

FORMER PERIOD.

765,000 tons of coals, at 4*s.* per ton, ...	£.153,000
Lime, ...	12,500
Pig iron 3600 ton, at 6*l.* 10*s.* per ton, ...	23,400
Increased value by a part re-cast into utensils, ...	14,600
35,000 bars lead, at 1*l.* per bar, ...	35,000
	£.238,500

AT

AT PRESENT.

1,250,000 tons of coals, at 5s. per ton, ...	£.312,500
Lime, ...	14,000
9500 tons of iron, at 7l. per ton, ...	66,500
Increased value by re-casting, ...	28,000
35,000 bars of lead, at 2l. 2s. per bar, ...	73,500
	£.494,500

SECT. VI.—WATER.

THERE are a great many lakes in different parts of the county, none of which are so remarkable for extent, or any circumstance attending them, as to merit a particular description; and an enumeration of the whole would be tedious and uninteresting. All of them contain some kinds of fish, such as trout, pike, or perch, &c. and are sometimes resorted to by anglers; but the quantity of food thus obtained is inconsiderable.

The Clyde, and its tributary streams, make the principal figure under the article of water. The main source of the river, rises in the ridge of mountains which separate the county from that of Dumfries. It is there called the Daer, and flows for several miles under that name, till it is joined by a little brook, called the Clyde, and from thence downward, has always the name of the Clyde. The principal streams by which it is joined in its course, are the Deninten, the Douglas, the Nethan, the Avon, and the West Calder, on the south side; the Medwain, the Mouse, the South and North Calders, and the Kelvin, which divides the county from Dumbartonshire, on the north side.

The river is navigable only to Glasgow. After which,

besides

besides a number of corn and other mills, it drives the machinery of two large cotton spinning works, at Lanark and Blantire. In great rains, particularly those which come from the S. E., it is sometimes swoln to a great height, and does considerable damage, more especially in Autumn, when the crops on the valleys by the river side are swept away, or much injured. The highest land flood remembered, was that of the 12th of March, 1782, when the river rose from 16 to 24 feet above the level of low water, according to the expansion or contraction of the banks in different places. In some places, the valleys have lately been fenced against inundations, by sloping banks of earth faced with grassy turf.

Most kinds of fish, which are found in the other rivers in Scotland, are also found in the Clyde, particularly the salmon. But though the river, from the foot of the lowest fall to Glasgow, runs, for about 20 miles, mostly on a bed of fine gravel, abounding with shoals and proper spawning places for that fish, and the small rivers also affording many places for the same purpose, there are perhaps few rivers of the same size less stocked with salmon. This is probably owing to the populousness of the country. Since the more numerous the population, there is likely to be the more people, in proportion to the number of fish, disposed to destroy them, at improper seasons, and consequently the destruction is greater. The fish are not only improperly and wastefully destroyed at the time of spawning, when many thousands perish at one blow, but through the Spring and Summer, numbers of thoughtless people swarm along the sides of the river, angling, and kill more fry, at three or four inches long, in one day, than all the grown fish caught in a season. It is probable that, if this waste could be checked, the river would soon become a considerable source

of

of food and revenue to the inhabitants. The proprietors of the fisheries on the Tweed, have entered into an association for improving them, by preventing all fishing at improper seasons, which, it is said, has had a very good effect. If all concerned in the fisheries of the Clyde would concur in adopting similar measures, and not only put a stop to all illegal fishing, but abstain themselves, for a year or two, from the most severe use of their rights, such as taking salmon in draught nets and cruives, it is probable the river would yield twenty times the quantity of fish which it does at present.

CHAP. II.

STATE OF PROPERTY.

THE valuation of the yearly rent of this county, in Scots money, established by the treaty of Union between the two ancient kingdoms, as a rule by which the land-tax and other assessments are proportioned, is as follows:

		£.	s.	d.
1	Carluke parish,	6001	14	3
2	Lanark,	4217	9	10
3	Carstairs,	2150	0	0

4 Carnwath,

		£.	s.	d.
4	Carnwath,	4978	19	4
5	Dunsire,	1450	0	0
6	Dolphinton,	850	0	0
7	Walston,	1233	0	0
8	Biggar,	3323	6	6
9	Liberton,	2501	8	0
10	Lamington,	2600	0	0
11	Coulter,	1600	0	2
12	Crawford,	5813	11	4
13	Crawfordjohn,	2360	6	8
14	Douglas,	5100	6	5
15	Roberton and Wiston, united,	2066	6	8
16	Simontoun,	838	0	0
17	Covington,	1333	0	0
18	Pettinain,	1570	0	8
19	Carmichael,	2246	7	4
20	Lesmahagow,	9907	0	0
21	Dalserf,	3820	0	0
22	Stonehouse,	2721	1	4
23	Glasford,	2653	3	10
24	Avondale,	7656	14	2
25	Hamilton,	9389	7	5
26	Blantyre,	1684	11	4
27	Kilbride,	7679	1	10
28	Shotts,	6558	0	10
29	Dalziel,	1232	19	10
30	Cambusnethan,	5400	2	0
31	Bothwel,	7389	16	4
32	Cambuslang,	3235	17	10
33	Old Monkland,	6481	9	10
34	New Monkland,	6821	18	4
35	Rutherglen,	1200	0	0
36	Part of Cathcart,	925	0	0

37 Car-

	£.	s.	d.
37 Carmunnock, ...	1650	10	0
38 Govan, ...	4702	18	7
39 Barony of Glasgow, ...	13002	9	6
40 Calder, or Cadder, ...	6272	16	8

Total, £.162,118 16 10

SECT. I.——ESTATES, AND THEIR MANAGEMENT.

THE following situation of land property was made up from the county books in the year 1794, and though a good deal of land in the county has changed its proprietors since that period, these alterations do not appear to be so material as to demand a new statement.

1st, The most considerable land proprietors, who hold estates, the valuation of the least of which is upwards of 2000l., are 11 in number, and the amount of their valuation is £.48,374 0 0

2d, The proprietors, who hold estates valued from 2000l. to 1000l., are 15 in number, and the amount of their valuation is 19,433 3 8

3d, There are 39 proprietors, who hold estates valued at between 1000l. and 400l., and their valuation amounts to 24,598 11 0

4th, There are 138 proprietors, holding lands valued from 400l. to 100l., whose valuation amounts to 24,008 9 6

Carry over, £.116,414 4 2

Brought

Brought forward,   £.116,414 4 2

5th, Those who hold properties below the valuation of 100l. are very
numerous, being near to 900, and
the whole of their valuation is 35,652 6 0

6th, The lands belonging to burghs,
and other societies and bodies corporate, or dedicated to the support
of hospitals, &c., the valuation of
which amounts to 10,052 6 8

Total,   £.162,118 16 10

The lands comprehended under the three first classes
of the above enumeration, are generally either the property of families of a considerable standing in the county,
or are parts of the estates of some of the great landholders
of the neighbouring counties, and being for the most part
under entail, remain in the possession of the same race of
proprietors, while that race continues. This, however,
is not universally the case : some are free of entail, and
lands to a considerable value, included in these classes,
have been lately in the market.

The lands comprehended in the fourth and fifth classes,
are more seldom under entail, and do not often continue
many ages in the same family ; particularly in the more
populous parts of the county, where the wealth gained
by commerce or manufacture is frequently laid out in
the purchase of land ; and the buyer and seller exchange
employments.

Management.—Upwards of three-fourths of the surface
extent of the county is the property of great landholders.
Those who have the greatest part of their estates within
this county, have country residences in it, which they
generally

generally occupy at least for a part of the year. Part of the lands round their habitations is cultivated under their direction ; and much of it has been greatly improved, in the course of the last 40 years. Many have extended their improvements still wider, especially in sheltering and adorning their estates, by inclosing and planting. One gentleman, ANDREW STIRLING, Esq. of Drumpellier, who, by purchases mostly recent, has become a great landholder, has highly distinguished himself as an improver. Bred in the commercial line, he has carried the enterprising spirit of that profession into all his transactions as a landholder ; and though his purchases, probably, were chiefly made with a view to coal, in the discovery and working of which he has been very successful, he has been no less attentive to the improvement of the surface, and, by a well conducted course of industry, has given such an addition of fertility to an extensive tract, that a person who had not seen it for 20 years, if brought on it now, would scarcely believe it was the same country*. Others might be signalized on the same account ; such as Major Gen. Sir. JAMES STEWART DENHAM, of Coltness, who, by uncommon exertions, has made valuable improvements on a considerable extent of land, naturally very unpromising. The same may be said of WALTER CAMPBELL, of Shawfield, Esq. But the business of this work is to give a general description of the state of the county, rather than to celebrate the conduct of individuals ; and it is hoped, that what is said of the gentlemen mentioned, will not be supposed to depreciate the merits of those who are not.

A considerable extent of enclosed land on the different

* These highly improved lands are now in the market, and will, no doubt, bring a great price.

estates is kept mostly in grass, and let out from year to year in pasture, it being broken up only for a few crops of corn, at distant periods, and again sown out with grass seeds. But the greatest part is rented upon leases to husbandmen; and the right of possession is secured to them and their heirs by an old statute, although the land should change its proprietor, so long as the lease lasts.

The proprietors of small estates, such as those of the fifth, and many of those of the fourth class, frequently cultivate the whole, or a considerable part of their own lands; and much of the improvement of the county is owing to their efforts and example; so that a person passing through it, has no difficulty in distinguishing those parts which are divided into small properties, from those which are not.

SECT. II.—TENURES.

THERE is no instance, known to the Writer, of lands being held, in this county, by any other tenure than that which is common to Scotland, by the custom of which the feudal system is still followed in the conveyance of landed property. According to this system, the sovereign is understood to be the sole lord of all the soil, minerals, and waters, possessed by his subjects. From him alone, therefore, the legal right of enjoying the property of land is supposed to be derived; and certain annual acknowledgments are due to him in return*. This right must be renewed upon every succession, whe-

* Some lands are held of the heir-apparent of the crown, but the nature of the tenure is the same.

ther of an heir or a purchaser, and a fixed fine paid.
The landholder, thus invested by royal charter, is possessed of a certain emanation of the same sovereignty,
proportioned to the extent of his domains*, and can
give charters to others, to hold such parts of them of
him, in the same manner as he holds of the sovereign,
and upon such conditions as he thinks proper; and these
again, can parcel out what they have thus acquired, in the
same manner; and so on. But it ought here to be understood, that when lands are sold, the purchaser may
either hold of the seller, or of the person of whom the
seller held, according to the agreement between the
parties.

In all cases, he who conveys is called the *superior*, and
those who receive the conveyance the *vassals*. The annual quit-rent paid by the vassal to the superior, is called
feu; and the fines upon successions, *casualties*. Lands
held by charters immediately from the sovereign, are
called *freeholds*; and those conveyed by subject superiors,
base holdings. Those only who hold their lands by the
first kind of tenure, can be electors of a representative
of the county in the House of Commons; and except only
in a few instances of charters of an old date, to which
this right is specially annexed†, the extent of valuation
must be 400*l.*, to give the qualification for this privilege.
As noblemen, who, either sitting or being represented in
the Upper House, are excluded from interfering in the

* It having been supposed that the authority which the barons had
over their vassals, had, in the rebellion 1745, brought numbers to the
field against government, contrary to their inclination, an Act of Parliament passed in the year 1748, which, without overturning the feudal
system, broke its force, and reduced this deputed sovereignty to a
shadow.

† Called Retours on a forty shillings land of old extent.

elections

elections of the Commons, hold lands in this county, to
the amount of 34,000*l.* of valuation, and as those of the
fourth, fifth, and sixth classes, with all the base holdings
which may be in the three first classes, are also out of the
question, neither the number of electors, nor the land
they represent, can be considerable.

The unwieldy and artificial manner of securing and
transferring landed property, according to the feudal
system, as above described; the frequent renewal and repetition of deeds and writings of different names and
distinctions; and all the various formalities requisite,
have rendered the aid of men learned in the law indispensable. The employment of these increasing as the
commerce and wealth of the country increased, has occasioned an increase of their numbers: and the increase of
numbers, again, has whetted their ingenuity to enlarge
the employment, in which they have been abundantly
successful. They are now become a powerful corpora-
tion, whose head is in the capital, and its members extended over the kingdom. Having, by such means, accompanied with great professional habits of industry and
acuteness, become necessary on all occasions, they come
in with the proprietors and cultivators of land, for a
goodly share of the produce, without either directing the
improvements or assisting the labours of cultivation.

CHAP. III.

BUILDINGS.

———

THE buildings of a country, next to the land, are the least perishable wealth of the community; and while the latter furnishes support, the former affords shelter and accommodation to the inhabitants. It seems only in this view, that buildings can obtain a part in a work of this nature; for though the magnificence of the buildings shows the wealth and grandeur of a people, and the symmetry of the architecture the refinement of their taste, neither of these have any connexion with their skill and diligence in agriculture; nor is it to be expected, that the judgment of a plain agriculturist should be so much formed upon the models of Greece and Rome, as to give a critical description of them: and it is believed such a digression would not be expected, nor perhaps relished, by the most part of readers, if it were attempted. It will be sufficient, therefore, just to make mention of the different kinds of buildings; such as, 1st, Public buildings; 2d, The residences of the opulent; 3d, The houses occupied by those who are supported by their own labour; making, as we go along, what remarks may occur, which bear any relation to the cultivation of the country.

SECT. I.—PUBLIC BUILDINGS.

PUBLIC buildings again may be divided into three classes, according to the purposes for which they are intended; 1*st*, For instruction and consultation; 2*d*, For correction; 3*d*, For the reception of the diseased and indigent.—The most considerable of the first class, are the buildings of the University of Glasgow (in which, a correspondent regrets that there is no institution for agricultural instruction), and the different churches, &c. in the county. St. Mungo's church, in Glasgow, is the only edifice of the ancient Gothic architecture, which remains entire and in use. The other churches in the county, of an old standing, are generally very homely piles. The more modern ones are better built; and many of them being furnished with spires, besides their principal purpose, diversify the scenery, and heighten the beauty of the country.

With respect to the second class of buildings, the different prisons, as well as all the other prisons of Britain, were certainly intended principally for correction; but it has been much questioned how far they have answered this salutary purpose. A highly benevolent and enlightened character, the late celebrated Mr. HOWARD, stood forth, and recommended, as an amendment, to commit offenders to solitary confinement, where, in the privation of temptation and amusement, while they laboured for their support, serious thoughts might take place, and operate reformation. This has been tried in the city of Glasgow, an account of which is given in the following extract from Sir JOHN SINCLAIR's Statistical Account of Scotland, vol. v. page 514. " This institu-

E 3   " tion

" tion was begun in the year 1789, when, in order to try
" the effects of solitary confinement and labour, some
" buildings belonging to the city, and formerly used as
" granaries, were fitted up as separate cells, for the re-
" ception of persons guilty of crimes meriting such pu-
" nishment.   These have been gradually increased to
" the number of sixty-four, where the prisoners are kept
" separate from one another, and employed in such la-
" bour as they can perform, under the management of a
" keeper, and under the direction of a committee of
" council, who inquire into the keeper's management, &c.
" The members of the town council, also, in rotation,
" are appointed to visit, not only this, but the prisons
" and cells near the hospital, once every week, and re-
" port whatever appears to them to be proper either to
" be rectified or altered.   The keeper has a record of
" the sentences on which each prisoner is confined—
" keeps an exact account of the wages of their labour,
" and after defraying the expense of their maintenance,
" the surplus is paid to them, when the period of their
" confinement expires; and some have received from 5l.
" to 7l.   Experience in this, and other great towns
" where this institution has been established, has de-
" monstrated, that, of all the species of punishment for
" offenders of a certain description, solitary confinement
" and labour is not only the most humane, but the best
" calculated to answer one great end of punishment, the
" amendment of the offender."   The magistrates and
council, by the assistance of some beneficent donations,
have been enabled to erect a building more completely
adapted to the purpose.   It consists of six stories, 106
feet in length by 30 in breadth, and contains 126 cells
8 feet by 7 each.   There are also two projecting wings
of three stories each, for the accommodation of the
keepers,

keepers, and store-houses for holding materials, &c.—
Denholm's Hist. of Glasgow.

Institutions of this kind, on a smaller scale, established
in the different parishes, particularly of those parts of the
county, where the number of inhabitants is much increased by the mixture of people employed in the great
manufactories, would be much for the advantage of that
part of society who are engaged in the cultivation of
the soil ; and at the same time, if the experience above
mentioned may be relied on, be a great benefit to offending individuals, as well as to the manufactures in which
they were employed. There is no kind of property so
much exposed to pillage as that of the husbandman; and
by an unfortunate prejudice, which too generally prevails,
there is none, against the pillaging of which fewer scruples
are entertained. The prevalent manufactures give employment to numbers of young people, whose parents,
perhaps, were neither qualified nor disposed to store their
minds with moral instruction. These are, by the profit
of their labour, rendered independent, while they are incapable of the innocent enjoyment of freedom. By such,
the practice of the social duties is too frequently not only
neglected, but made the subject of sport and ridicule ; the
dissolution of manners thus extended, and idleness and
licentiousness kept in countenance and encouraged. The
violation of property, therefore, when it can be committed
with any chance of impunity, is little regarded, and the
produce of a neighbour's field seems almost to be thought
lawful prize : so that if the corruption of manners spread
with as great rapidity for a dozen of years to come, as it
has done in the dozen last past, it is difficult to conceive
how the husbandman can be protected in his rights, or
be allowed to pursue his employment with comfort. As
those giddy people, by whom the depredations on the

fields

fields are most commonly committed, by the idleness such practices induce, injure their own interest as much as that of others, it is the duty of all ranks to endeavour to put a stop to these flagitious practices. By the judicious use of a few cells, excluded from all communication, and from the view of all amusing objects, and provided with the implements of different artisans, erected in the most populous parishes, these depredations might perhaps be checked; and a number of thoughtless young people, not yet hardened in vice, whom a ferment of blood, unrestrained by proper habits, hurries into crimes, be brought to a sense of propriety, and become useful members of society.

The public is so little accustomed to sympathize with many of the hardships peculiar to husbandmen, that probably the offences above hinted, will be thought too trivial for these serious animadversions. But it will not surely be denied, that, as all are interested in the produce of the earth, which is committed to the charge of husbandmen, whatsoever tends to waste and destroy it, or interrupts and discourages the means used to augment its quantity, is of a nature the most generally injurious: and, therefore, the cultivators of land have at least as strong a claim to public protection as any other class. But supposing the cause of husbandmen out of the question, and any offence committed against them ever so trifling, the public should be reminded, that bad habits is the cause of great offence; and those who begin with plundering the fields, will scarcely stop there. An eminent French author observes, " Les mœurs sont non seulement le " tableau vivant de l'etat de la societé ; mais en sont " encore le ressort principal." And again, " Ou les " mœurs regnent, les loix les plus simple sufficent." Since manners are of such importance, it is surely proper to begin in time to correct them, where they are wrong. That the advancement of manufacture, while it promotes the prosperity, destroys the morals of a people, is but a vague indiscriminating observation, calculated to inspire a groundless despair and indifference for the concerns of the public. That superfluity of wealth, which has been considered as one great origin of dissolute manners, cannot descend to the great mass of the people; nor can the exercise of any branch of industry, of itself, corrupt the morals of the person who earns his livelihood by it. The visible declension of morals cannot, therefore, be solely attributed to the advancement of manufacture: nor does it appear to be a very romantic hope, that, if the attention to the subject were equal to the interest which people of all ranks have in it, such a degree of decency and propriety of manners might long be preserved, among the body of the people, as the comfortable existence of society absolutely requires: and surely no means suited to accomplish so valuable an end, ought to be neglected.

Of the third class of public buildings there are several in this county, both for

"The young who labour, and the old who rest;"

an enumeration of which, it is believed, would be needless. What is called the Poor's House, in Glasgow, and the Infirmary lately built there by public contribution, for the reception and cure of the sick and wounded, are the most prominent. Were it convenient here, to give an account of the charitable contributions in favour of these institutions—of the attention paid, and the wise measures pursued, by societies and individuals, in conducting them—of the numbers of helpless and deseased people entertained, and the benefits they have received, it would do honour to the humanity and liberality of the inhabitants,

inhabitants, particularly those of the present age, by whom so many additions and improvements have been made.

SECT. II.——RESIDENCES OF THE OPULENT.

THE citizens of Glasgow have long displayed a great taste for elegance and magnificence, both in their public and private buildings; which has still been extending as wealth increased; and has been exerted, of late years, with almost unremitting activity. There are now perhaps few towns in Europe, which can boast superiority to Glasgow, in the spaciousness of its streets, the regularity of the plan, and the symmetry and magnificence of the buildings. The amount of the rent of houses in Glasgow has been estimated at 100,000*l.* and upwards.

Besides the large buildings in towns for the accommodation of the opulent, excepting in the most elevated parts of the county, where the situation is too bleak and forbidding, the face of the country is finely interspersed with the seats of the principal landholders, or the villas of the wealthy merchants, manufacturers, &c. The attention which has been generally paid to shelter and adorn the grounds around these, has contributed greatly to the beauty, and also to the fertility, of the country; gardening, purely ornamental, having hitherto abstracted but little land from the main use of producing food either by corn or pasturage.

SECT. III.—THE HOUSES OCCUPIED BY THOSE WHO ARE SUPPORTED BY THEIR OWN LABOUR.

AMONG these may be ranked the houses of the proprietors of the fifth class, in the enumeration contained
in the last chapter, and a part of those of the fourth: for
though the occupants of these derive a part of their income from their land rent, they depend still more on
their own industry in the cultivation. The most part of
these do not differ greatly from the better sort of farm
houses occupied by tenants. The farm houses and offices have been much improved of late years: in general,
however, they are not so comfortable, nor so well adapted
to all the purposes of agricultural improvement, as they
ought to be. It is not necessary, indeed, here, to have
the offices so extensive as in places where the crop is
stored within doors; it being the general opinion, that
both corn and hay are best preserved in the open air;
and for that reason, no hay, and only that part of the
corn which is intended to be first thrashed out, is stored
in the barn; but there is a great want of sheds and convenient straw-yards for young cattle, &c. The high
price of slates, and the distant land-carriage to many parts
of the county, has much discouraged the use of them,
which is also a disadvantage to husbandry, as thatched
roofs give great harbour to vermin, and the covering of
so many of them, at the end of every short period, with
straw, consumes a great deal of what would be better
bestowed in littering live stock, and making manure, an
article which, important as it is, cannot be otherwise
procured in many places distant from towns. There are,
it is true, many tiles manufactured and used in the county;
but when that manufacture was first begun, the tiles were

of

of a bad quality, and those who used them had cause to repent it, which deterred others from following the example; so that, though great improvement has been since made in the manufacture of tiles, they are not yet very generally used for covering farm houses. As convenient accommodation is better understood, and more eagerly desired than heretofore, and as some of the most considerable landholders are disposed to make such additions and amendments to the farm houses as shall be required, on reasonable terms, it may be expected that, as leases fall, much improvement will be made in this article.

It is in vain to say any thing of the ancient cottages of the county, the former nurseries of field labourers; for they may be said to be now no more; as the few scattered ones which still remain, can scarcely be called an exception. It having, for a long time, been the custom of this county for farmers to keep only unmarried servants, who are lodged and fed in the house, for the execution of agricultural labour, the cottages on the different farms have dropt gradually into ruins, and been removed; and the small tenements being mostly swallowed up in the larger farms, the cottagers and the farm servants, when they marry and settle, withdraw from their rural habitations to towns and larger villages, to which the increase of employment also invites them; and their progeny, who, formerly were from their infancy habituated to the labours of the field, are mostly occupied in some branch of manufacture: so that the means by which the necessary supply of labourers in husbandry used to be obtained, is in some measure cut off.

This change hath been pathetically bewailed by persons of feeling hearts and warm imaginations, who, being charmed with the simplicity of rural life, have painted the sequestered cottages as the calm retreats of innocence

and

and virtue; while the disciples of a more rigid school have justified and extolled the measure of driving the superfluous inhabitants from the country, to follow industry in towns. It is probable that neither the former had very accurately considered the subject of their lamentations, nor the latter made proper calculations how many inhabitants any district of the country ought to contain, before it became necessary to drive any of them away. The diminution of cottages in this county does not, indeed, appear to have proceeded from any premeditated plan of economy, but from fortuitous causes. But certain it is, whether the scarcity of field labourers be ascribed chiefly to the prospect of superior ease and comfort, to which a growing manufacture invites, or partly to the little attention paid to preserve those in their former habits and situation, who might have been willing to remain, it has been much felt for some years past, and still seems to be increasing. No doubt, the war, which has carried the flower of our peasantry to the army, has also contributed. Many can never return; those who do, will hardly bring all their former industry and activity along with them. But if the capital employed in agriculture were equal to what the complete cultivation of the county would require, many more hands would be wanted, while, in the mean time, the former means by which the usual recruit was obtained, is cut off by the demolition of cottages; and it still remains to be seen, whether or not Dr. GOLDSMITH's famous distich,

" But a bold peasantry, their country's pride,
" When once destroy'd, can never be supply'd,"

be as just as it is poetical.

The county, however, is supplied with a new set of cottages. Several landholders, partly perhaps to prevent the depopulation of the country, and partly for their

own

own emolument, have lett out, either in feu or long leases, spots of ground, for houses and little gardens, generally upon the sides of the public roads. Upon these, many little handsome cabins have been erected, which, accompanied with neatly dressed gardens, supplied with pot-herbs, and frequently ornamented with a few flowers, have a very pleasant effect. These are mostly clustered into villages, some of which are become pretty populous. But though many of them are occupied by the inhabitants of the old cottages, or by farm servants when they marry, most of the children betake themselves to some kind of manufacture; and, in a softer life, lose both the relish for, and the power of performing, the more rugged labours of the field.

Farm houses are built at first by the landlord, and the tenant is bound to keep them in repair during the time of his lease, and to leave them in good condition at the expiration of it. Every successor is bound in the same manner. Of late, some landlords contract to make the farm houses of a certain money value at the commencement of the lease, and the tenant to leave them of the same value at the end of it, or to pay the deficiency, as the same shall be estimated by men of judgment: on the other hand, if the tenant, for his own conveniency, shall have made the houses of greater value than they were at the first, the landlord pays to him the increased value. The proprietors of cottages bear the expense of all the necessary repairs (glass excepted), unless there be a particular agreement to the contrary.

CHAP. IV.

MODE OF OCCUPATION.

————

THE face of this county being greatly diversified, the mode of occupation is different in different parts. The mountainous district, at the head of it, is occupied mostly with flocks of sheep: upon the ridges on the E. and W. sides, where the ground is marshy, and less proper for sheep, and the exposure too bleak to encourage the cultivation of corn, cattle are mostly pastured, and those generally milch cows and their young, many of which are reared; a small quantity of corn only being cultivated, principally for the sake of winter provender: the less rugged and less exposed parts are more occupied in the culture of corn, &c.

————

SECT. I.—SIZE OF FARMS.—CHARACTER OF FARMERS.

EVERY parish was formerly divided into ploughgates, each of which consisted of from 70 to 120 acres of arable land. One of these, for the most part, made one farm. In some cases, a ploughgate was divided between two farmers; and there were sometimes small tenements annexed to ploughgates, not more than a fourth, or, as it

was

was called, a horse-gang, having a house, and occupied
by a sub-tenant. The greatest part of the farms are still
moderate, renting from 30*l.* to 150*l.* yearly; but of late
some farmers have obtained larger possessions, some renting from 200*l.* to 600*l.* In the sheep pastures, the farms
are very extensive.

The husbandmen of this county are hardy, active, and
laborious, well qualified to struggle with the difficulties of
soil and climate above described, and equally frugal and
economical. To use the words of a late celebrated
satirist, they are " not overburthened with unwieldy
" knowledge." Instead of pretending to tear away the
veil, under which Nature has concealed many of her important operations, they are attentive to acquire experience, by observing what passes under their notice.
Having now, in a great measure, shaken off those fetters
in which prejudice holds husbandmen, perhaps, faster and
longer than any other set of men, they communicate
with one another, and mutually learn such practices as
tend to improvement. Such in general is their professional character.

Their moral character is, probably, nearly the same
with that of the inhabitants of other districts in similar
circumstances. Men of acute feelings who, in their commerce with the busy world, have been hurt with the in-
sincerity and knavery of mankind, contrasting the open
unstudied address of countrymen with the artificial manners practised in the crowded walks of life, have represented the country as the sole abode of innocence, and
countrymen as the most virtuous of the human race.
This fair portrait of rural manners, however, has not
been suffered to pass undebauched. Much ingenuity has
been employed, and much pretended sagacity displayed,
in drawing aside the countryman's garb of simplicity,
and

and showing the craft, hypocrisy, sordid avarice, &c. supposed to be concealed under it. The failings of humanity, no doubt, will accompany man, in whatever situation he is placed: but when it is considered in how little estimation husbandmen are held, it is less a wonder, that vices are to be found among them, than that there are so few. The spruce citizens laugh at their rustic, homely appearance, fly in a rage when the price of country commodities are raised in proportion to the demand, and enter into combinations to beat them down. If farmers become rich by their industry and good fortune, they are accused of extortion, of storing up their corn, or sending it to foreign markets, to starve the poor;—if they are unfortunate, they are despised. The haughty lordling, who should be their patron and protector, regards them only as an inferior race, formed to toil for his enjoyment; and the family who have run a lease of industry and good management, are dismissed at the end of it, if a stranger offers a little more rent. Against such an host of depression, the insulated husbandman has no shelter, but silent dissimulation, and it is not surprising that he should sometimes have recourse to it.

This stricture is not meant to extend to all. The conduct of the rational and the discerning of every rank, will be very different. But still such prejudices are too common; and the cultivators of the country do not seem to possess that respectability which the importance of their situation deserves. It is an observation not less true than trite, that self-respect is one of the best guards against the commission of crimes; and that people will be more disposed to respect themselves, when they are respected by the world. It is a generous employment that of a husbandman: let it be regarded as such, and the virtues

CLYDESDALE.]　　　　F　　　　congenial

congenial to it will be more conspicuous among its professors, and render them still more eminently useful to society.

The political character of husbandmen, one would have thought, should have made them the darlings of statesmen and politicians. Strongly attached to their native soil, and fully occupied in the diversified employment they find upon it, they neither have the disposition nor the leisure to enter into speculative disquisitions concerning government: and while the people engaged in manufacture are perpetually lifting up the heel against that administration which has favoured the sources of their support, neglected husbandmen remain peaceful and passive, except in extraordinary emergencies, when they stand forth to offer government such feeble assistance as they can bestow. The truth of this position will be best illustrated by studying the history of mobs, by which it will appear, that a thousand of these have risen in towns and populous manufacturing places, for one in the country; and in the few which have happened in the latter, scarcely any person has been concerned, who deserved the name of cultivator, or was possessed of the habits naturally attached to that employment. *Pius quæstus, stabilissimusque, minimeque invidiosus; minimeque male cogitantes sunt, qui in eo studio occupati sunt.*

The religious character of the husbandmen of this county, and, to speak more largely, of the generality of those in the kingdom, is a zealous adherence to the form of church government, and mode of religious worship, established by their ancestors at the Reformation. Many, indeed, have withdrawn from the church, as now established; but it is because they apprehend that church had swerved from its original purity. The support of all the teachers, &c. of the different sects, has greatly

increased

increased the expense of public instruction, a considerable part of which is drawn from the cultivators of land: and it has been much doubted, if the cause of real religion has been advanced in the same proportion. It is not convenient here to investigate the causes which have produced those religious schisms; but it may be justly regretted, that, amidst the various controversies upon abstruse and speculative points, the sublime morality, inculcated by the christian religion, so consonant to the state in which husbandmen are placed, and so well adapted to afford them consolation in all difficulties, should be so slightly regarded.

SECT. II.—RENT.

THE rent of land, in this county, is mostly paid in money, the old personal services, so oppressive to the farmers, and so unprofitable to landlords, being generally abolished. A few fowls are still exacted from the farmers of some estates, besides money rent; and the restriction to carry the corn to certain mills to be ground, is still continued. But by this, oats only is meant, which was and is still considered as the principal bread corn. This is a real grievance, not so much on account of the heavy exaction levied for the grinding, which is generally double, and sometimes triple, of what would be required, at a mill to which the farm was not bound, as the farmer, knowing the extent of this exaction, may be supposed to have laid his account with it, but being bound to carry the corn to be ground by a person, perhaps, neither whose skill nor honesty can be depended on, and from whom little civility is to be expected. Farmers bound to a mill, are also bound to carry mill-stones from the quarry, whatever be the distance, and to assist at repairing the mill, mill-dam, and mill-lead, when required. These are absurd servitudes, and ought to be abolished. And since this Report was last published, a law has passed for relief of this hardship. Some farmers are bound to pay the land-tax and other public burthens; but these payments are oftener made by the proprietor.

As some parts of the county differ widely from others, with respect to fertility, so also the rent of land is not less different. Mountain and moor pasture is not lett by the acre, but by the number of sheep the pasture will support, the healthiness of the ground, the feeding quality of its grasses, &c. Superior grounds, lett in pasture, have for a good many years been rented at from 3*l.* to 10*s.* per acre, according to the quality of the pasture; and rich pasture ground has now gone as high as from 4*l.* to 5*l.* per acre. Land of a deep rich soil, which has long been improving in fertility by remaining in pasture, has, for several years past, lett for two or three corn crops, from 7*l.* to 11*l.* per acre per annum: and this present year, 1805, there are instances of land lett in that way at 12*l.* 12*s.* per acre. In one instance, there is land perhaps near 600 feet above the level of the sea, lett for the ensuing crop at 12*l.* Before the year 1760, the rent of land in this county was generally very low. It began, about that time, to make rapid advances, as leases expired. Yet within these 20 years, there were many farms in which the average rent of the whole arable lands was considerably under 10*s.* per acre. Of late, the rise of rent has been excessive. As leases have fallen, the rent has been raised from one and three-fifths to triple the former rent: and it is believed, that upon an average, the rent of land is considerably more than doubled.

doubled. Unfortunately, the rise of rent has all along pre-
ceded the accumulation of agricultural capital, so that,
except in a few cases, where circumstances uncommonly
lucky have occurred, the farmers are not wealthy, and
consequently their efforts to improve the soil are feebled.
It is difficult to account for the great rise of land-rent.
The great advance in the price of live stock, and of the
produce of the dairy, indeed, justifies a considerable rise
on the rent of farms chiefly adapted to pasture. But
how shall we account for the excessive rise on tillage
farms? The price of all kinds of farm labour, and of
every article needed in a farm, is more than doubled
since the year 1790. The farmer's household and personal expense, is also considerably increased. But the increase in the quantity and value of disposable farm produce, bears no proportion. The high prices of corn
which have repeatedly occurred of late years, enriched a
few farmers, in favourable circumstances. But a great
many had no corn to sell; and were sufferers in the general calamity. An Act lately passed in Parliament, has
been supposed by some to be a scheme of the movers, to
keep the price of corn always high: and whatever was
the design, it is evident that, in a country like Britain,
which does not produce corn sufficient to feed its inhabitants, every thing which cramps the freedom of the
commerce in corn, must have that effect. But every
lover of his country will deprecate the success of any
such scheme. Notwithstanding the liberal reward which
attends almost every species of labour in Great Britain,
there are great numbers of the labouring poor, such as
the old, the maimed, the valetudinary, and those who
are loaded with a number of young, and often sickly
children, whose earnings can scarcely purchase them any
other food than meal and potatoes. Those people are

F 3 always

always thrown into some degree of distress, whenever the price of grain is much higher than they are accustomed to pay for it in years of ordinary plenty. But if, by a monopoly in favour of the growers of British corn, its price should be frequently put beyond the abilities of the poorer sort to purchase their subsistence, they can no longer live by their labour. They will then sink into despondency. Finding that their labour cannot procure them bread, they will lose that honest pride which prompts people to depend on their own industry for support; they will prefer idleness, and have recourse to eleemosynary aid. Thus many whose labour contributed somewhat to the increase of the national wealth, will become an useless burthen; and a monopoly to raise the price of corn, instead of advancing the rent of the landholder, will diminish it, by increasing the burthen of poor's rates. We may conclude, therefore, upon the whole, that the rent of land is too high, since there are few instances, of farmers improving their stock at the same rate as daily happens in other employments; and since it has been found necessary to keep up that rent by monopolies, which may endanger the public prosperity.

It is impossible to give an exact state of the whole land-rent of the county, many of the statistical accounts of the parishes being silent on that head. In the report made 1794, it was attempted to give a general idea of it, and though it was in part conjectural, it was probably not far from the truth. The same was given in 1798, allowing the reader to make what alteration he thought proper for the difference of the times.

In the upper ward, the sheep pastures of Crawford will maintain a sheep for every two acres, and the rent paid per head is about 3s. In some of the more wet

and

and barren moors, it will perhaps require three acres to maintain each sheep, and a farmer cannot afford to pay more than 1s. 6d. per head. The arable land, fertile as much of it is, with all the inconveniences of climate, distance from market, and the many spots of poor land intermixed, which cannot be profitably improved, on account of the scarcity of manure, and other discouragements, does not perhaps yield of rent, on an average, above 8s. per acre. The moorish land of the middle ward, when the mosses above described are included, is still less valuable than that of the upper ward. The arable land, though from the circumstances of its situation, and the great improvements which have lately been made, some of it letts, at times, nearly as high as most land in Britain; yet there is a great proportion so sterile in its nature, and so unhappy in its exposure, that the average rent of the whole probably does not exceed 14s. per acre. The under ward, though not originally more fertile, on account of its situation in the neighbourhood of Glasgow, is now more valuable. The average rent may be computed at 25s. per acre. The following is a kind of scheme of the county:

UPPER WARD.	Acres.	Acres.
Moor pasture,	185,000	
Woods,	3,140	
Channels of rivers, brooks, roads, &c.	2,060	
Orchards,	70	
Arable and meadow,	76,490	
		266,760

MIDDLE

	Acres.	*Acres.*
MIDDLE WARD.		
Brought forward,		266,760
Moors and coarse pasture, ...	66,000	
Woods, ...	4,150	
Channels of rivers, sites of towns		
and villages, roads, &c.	1,300	
Orchards, ...	130	
Arable, ...	70,750	
		142,330
LOWER WARD.		
Woods and waste ground, ...	1,000	
Sites of towns, roads, &c.	1,500	
Arable, ...	33,850	
		36,350
Total acres, ...		445,440

The yearly rent may stand thus:

		£.	s.	d.
UPPER WARD.				
Pasture, 185,000 acres, at 1s. £. 9,250				
Arable, 76,490 acres, at 8s. 30,596				
	39,846	0		
MIDDLE WARD.				
Pasture, 66,000 acres, at 6d. 1,650				
Arable, 70,750 acres, at 14s. 49,525				
	51,175	0	0	
LOWER WARD.				
Arable, 33,850 acres, at 25s.	42,312	10	0	
Total rental, ... £. 133,333		10	0	

This scheme has been judged, by some intelligent people who have considered it, to have come very near the truth
at the time it was made. It is believed that the rent of
land

land is more than doubled since, but supposing it to be double, the rental of the county will be 266,667*l.*

SECT. III.——TITHES.

THERE being nothing, relating to tithes, peculiar to this county, this article might be immediately dismissed. But as these Reports fall under the perusal of readers unacquainted with the customs of Scotland, it seems necessary that these customs should be explained, since this has not been done in the Mid-Lothian Report (the only one which the Writer of this has seen), in which it was most likely to be expected. It was for this reason, that, in a former chapter, an attempt was made to explain the nature of Scots tenures; and, for the same reason, it may be proper to do the same with any similar article which may occur.

In Scotland, as well as the other countries of Europe, the Jewish law, Numb. xviii. 21.*, was adopted, in making provision for the ministers of religion: but the tenth part of the produce of a barren land, was little for supporting the numbers whom superstition, in the days of ignorance and idleness, induced to embrace holy orders. The clergy, however, in the influence which their sacred office gave them over the minds of the people, possessed, in superstitious times, abundant means of increasing their funds, of which they knew well how to avail themselves. While they directed others in the way to Heaven, they accepted the reward of their pious

---

* And behold, I have given the children of Levi all the tenth in Israel, for an inheritance, for their service which they shall serve, even the service of the tabernacle of the congregation.

labours

labours in the goods of this world; and of these they
had at last obtained a large share. Dr. ROBERTSON, in
his History of Scotland, says, " The Scottish clergy paid
" one half of every tax imposed upon land; and as there
" is no reason to think that, in that age, they would be
" loaded with an unequal share of the burthen, we may
" conclude, that, by the time of the Reformation, little
" less than one half of the property of the nation had
" fallen into the hands of a society, which is always
" acquiring, and can never lose." When the dawn of
knowledge had somewhat dispelled the mist of superstition, and the bold truths every where advanced, had
battered down the bulwark of sanctity, with which the
clergy and their possessions were surrounded, their
wealth became a tempting prize to the avidity of the
laity, and they were hunted down for the sake of the
spoils they possessed. The noblesse and retainers of the
court, not only got hold of the greater part of their
monasteries, castles, manors, and extensive landed property, but obtained grants from the crown, of the tithes
which had been originally destined for the support of
religious institutions, with the reservation only of so
much as should be necessary for the maintenance of a
spiritual pastor in every parish; and a very little was,
in those days, thought sufficient for this purpose. The
other land proprietors, many of whom had been equally
active in pulling down popery, complained loudly of this
unequal distribution of the spoil. CHARLES I., who then
sat on the throne, unable to govern, in those turbulent
times, was willing to temporize; and having accepted the
office of arbiter between the contending parties, decreed,
that all those who had obtained grants of the tithes of
any district, who were generally called titulars of teinds,
should be obliged to sell, to the proprietors of the land,
                                              a right

a right to as much of the tithes (commonly called teinds
in Scotland), as was not exhausted in the maintenance of
the parochial pastor, at the rate of nine years produce:
and the proprietors who wished to purchase, were authorized to institute a suit before a court, which had
been erected for such purposes in the preceding reign,
to obtain a judicial valuation of these teinds, and operate
a sale*. Buildings and orchards are not comprehended
in these valuations; and the proprietor is allowed a de-
duction for the expense he has recently laid out in improvements, such as enclosing, draining, &c.: so that it
is only from negligence of proprietors, in not negotiat-
ing the purchases of teinds in time, that any tax has
fallen on the industry employed in improving the land.

The tithes being thus wrested from the clergy, cultivators are left at full liberty to increase the fertility of
the country, as far as possible, without being under the
apprehension, that those who have borne no part in the
culture should come in for a share in the harvest. But
the clergy, though no longer in possession of the teinds,
are not left to a precarious subsistence. That austerity
of manners and indifference about worldly enjoyments,
which distinguished the apostles of the Reformation in
Scotland, was soon laid aside; and the clergy began to
explain, that their temporal interest had been much disregarded, in the revolution which had taken place. A
court was therefore erected, as above said, the business
of which is now done by the members of the supreme
court, to take cognizance of their affairs: and to this
court the ministers have applied for augmentations of
their stipends, which has seldom been denied, when

---

* Instead of the tithes formerly levied in kind, a fifth part of the
land-rent was declared to be the teind.

there

there were any teinds remaining unbestowed in the parish. Each minister has, besides, at least four acres of arable land, called the *Glebe*, and so much for pasture, with a house, called the *Manse*, offices and garden.

At the last publication, a Table of the stipends of the established clergy in the county was given, chiefly extracted from Sir JOHN SINCLAIR's Statistical Account of Scotland, amounting to 6193*l.* 12*s.* 2*d.* But as several augmentations have been obtained since, and the rent of land and houses is considerably advanced, it was thought proper to insert the following, which, according to the information obtained, is nearly a state of the present value of the clerical livings in the county.

TABLE of the Value of the Livings of the Clergy in the different Parishes of the County.

PARISHES.	Money Stipend.			Victual, at 1s. 3d. per peck.			Manse & Glebe.			Total.		
	£.	s.	d.	Chal.	B.	P.	£.	s.	d.	£.	s.	d.
UPPER WARD.												
Biggar,	53	11	1	4	0	0	18	0	0	135	11	1
Carluke,	85	0	0	5	0	0	35	0	0	200	0	0
Carmichael,	72	4	5½	4	0	0	24	0	0	160	4	5½
Carnwath,	92	0	0	4	0	0	30	0	0	186	0	0
Carstairs,	55	0	0	6	0	0	36	0	0	187	0	0
Covington,	80	0	0	0	0	0	24	0	0	104	0	0
Crawford,	83	6	8	0	0	0	16	0	0	99	6	8
Crawfordjohn,	110	0	0	0	0	0	19	0	0	129	0	0
Coulter,	41	13	4	3	0	0	15	0	0	104	13	4
Dolphinton,	46	0	0	0	0	0	15	0	0	61	0	0
Dunsyre,	100	0	0	0	0	0	15	0	0	115	0	0
Douglas,	75	0	0	5	0	0	36	0	0	191	0	0
Lanark,	76	0	0	6	1	8	24	0	0	197	10	0
Lamington,	58	5	1	0	0	0	15	0	0	73	5	1
Lesmahagow, 1st minister,	44	13	6	9	0	0	40	0	0	228	13	4
Ditto, 2d minister,	55	0	0	6	0	0	22	0	0	173	0	0
Libberton,	110	0	0	0	0	0	30	0	0	130	0	0
Pettinain,	5	18	0	5	0	0	25	0	0	110	18	0
Simonton,	55	11	1	0	0	0	15	0	0	70	11	1
Walston,	33	0	6	3	0	0	15	0	0	96	0	6
Wiston and Roberton,	57	0	0	6	0	0	38	0	0	191	0	0
MIDDLE WARD.												
Avondale,	55	0	0	6	0	0	22	0	0	173	0	0
Ditto, assistant,	27	16	8	0	0	0	0	0	0	27	16	8
Blantire,	53	6	8	3	8	0	36	0	0	145	6	8
Bothwel,	46	18	0	7	1	0	36	0	0	195	18	0
Cambuslang,	35	0	0	11	8	0	30	0	0	249	0	0
Cambusnethan,	35	0	0	6	14	0	25	0	0	170	0	0
Dalserf,	60	0	0	6	0	0	30	0	0	186	0	0
Dalziel,	60	0	0	0	0	0	40	0	0	100	0	0
Glasford,	36	0	0	6	0	0	28	0	0	160	0	0
Hamilton, 1st minister,	41	13	4	12	0	0	42	0	0	275	13	4
Ditto, 2d minister,	44	3	4	9	0	0	20	0	0	208	3	4
Kilbryde,	0	0	0	15	0	0	30	0	0	270	0	0
Monkland East,	70	0	0	8	0	0	25	0	0	223	0	0
Monkland West,	70	0	0	8	0	0	30	0	0	228	0	0
Shotts,	140	0	0	0	0	0	30	0	0	170	0	0
Stonehouse,	16	12	6	6	1	7	24	0	0	138	1	3
UNDER WARD.												
Cadder,	0	0	0	8	0	0	25	0	0	153	0	0
Carmunnock, about	100	0	0	0	0	0	0	0	0	100	0	0
Cathcart, part of	40	0	0	0	0	0	0	0	0	40	0	0
Glasgow City, seven ministers, 250l. each,	1750	0	0	0	0	0	0	0	0	1750	0	0
High Church parish of ditto,	250	0	0	0	0	0	150	0	0	400	0	0
Barony of ditto,	250	0	0	0	0	0	130	0	0	380	0	0
Gorbals,	150	0	0	0	0	0	0	0	0	150	0	0
Govan,	50	0	0	8	0	0	40	0	0	218	0	0
Rutherglen,	0	0	0	9	3	14½	27	0	0	174	18	1½
	4770	14	2¼	196	5	13½	1317	0	0	9229	11	0¾
	£.3141	16	10½									

And the support for the teachers of the different sectaries, may perhaps amount to a fourth or fifth of the above. There are also many chapels of ease in the populous parts, well endowed.

———————

SECT. IV.——POOR'S RATES.

Poor's rates are imposed by an act of the Scots Privy Council, 11th August, 1692, which directs, that the heritors of a parish shall meet with the minister and members of the kirk session, who are jointly to make up a list of the indigent persons in the parish, and then impose an assessment for their maintenance, one half on heritors, in proportion to the valuation of their property, the other half on tenants and householders, according to their ability. Thus the people who pay the poor's rates, are both made the judges of the indigence of those who claim charity, and the imposers of the assessment for their relief; and this, no doubt, is the best guard which could be well devised, against an exorbitant assessment or a prodigal distribution.

In the populous manufacturing parishes, where the proportion of needy persons is always greatest, it has been necessary to have recourse to this mode of supporting the poor. In the more thinly inhabited parts, the poor have hitherto been supported by the interest of money which has been mortified by pious persons, by the offerings at the church door, of the people assembling to divine service on Sunday, by hiring out palls for funerals, &c. without having recourse to assessments.

SECT. V.—LEASES.

On sheep farms the leases are commonly short, as no process of improvement is carried on; on arable farms the most common length of lease is 19 years. In some cases, when the farmer undertakes extraordinary improvements, such as expensive draining, making and training up of fences, &c. the length of the lease is 31 years; and there are a few instances of still longer leases. The term at which a new tenant enters upon the possession of the land, is Martinmas, but the houses and pasturage are retained by the former possessor till Whitsunday. Of late, the old tenant is bound to relinquish to the new one, half the grass grounds, and lodging for labouring servants and horses, at Candlemas. In some estates, the farmers are bound to follow a certain system of husbandry and rotation of crops; in some they are bound only to have no more than a certain proportion, such as a half, a third, or a fourth, of the farm, in tillage crops annually; in others, they are left more to their own discretion; but, in all, they are laid under certain rules, meant to prevent them from exhausting the farm towards the expiration of the lease.

SECT. VI.—EXPENSE AND PROFIT.

It is not here intended to explore the books of expenditure kept by any of the husbandmen of the county. In general, the expense of every thing relating to the culture of land is become very high. A specimen is
here

here given, of the expense of sundry articles in rural economy.

Expense of preparing an Acre of Land for Turnips, in the upper parts of the County, 1804.

	£.	s.	d.
2 Ploughings and 2 ridgings, equal to one ploughing, with harrowing, at 8s. each, ..	1	4	0
25 Cart-loads of dung, at 3s. 6d., 4l. 7s. 6d.; carriage, spreading, &c. 8s. 6d.	4	16	0
Seed, sowing, rolling, &c.	0	3	6
First and second hand-hoeing, 4s. 8d.; 2 horse-hoeings 5s.	0	9	8
	£.6	13	2
If the land be full of the roots of weeds, it must have an additional ploughing, which with breaking, harrowing, gathering weeds, &c. will cost, ...	1	0	0
	£.7	13	2

In the middle and under wards of the county, the same operations will cost from a fourth to a fifth more. The preparation of an acre for barley (not succeeding turnips) is about the same.

Expense of preparing an Acre for Wheat in the dense soils of the Middle and Under Wards, by Summer Fallowing.

	£.	s.	d.
5 times ploughing, with harrowings, ...	2	10	0
40 cubic yards of dung, with carriage and spreading, ...	12	0	0
Seed and harrowing, ...	1	4	0
	£.15	14	0
		6 chalder	

Brought forward £. 15 14 0

6 chalder of lime is also frequently given,⎫
   which with carriage, slacking, and spread-⎬ 3 12 0
   ing, is about ....⎭

£.19 6 0

To this a year's rent of the ground while fallow, must be added.

But land is now less frequently summer-fallowed for wheat. It has been found to succeed pretty well after potatoes or beans, and sometimes on a clover stubble. The potatoes are planted in rows at 2 feet 3 or 4 inches distant, dunged, and kept clean by hand and horse hoe-ing. The land is also dunged for beans, and of late the beans are frequently sown in rows from 20 inches to 2 feet asunder, cleaned first with the hand-hoe, and af-terwards the earth is taken away from the sides of the rows by a little plough with one horse, and in a little time after returned again. The clover stubble is dung-ed. A moderate crop of potatoes or beans defrays the expense of manure, culture, &c. In all these cases, one ploughing serves for the reception of the wheat seed.

The cost of a full mounted cart is about 13l., of a plough 3l. 10s., harnessing a horse for cart and plough 3l. We might thus go on to state the cost of every thing relating to a farm, and with seeming accuracy make a computation of the annual expenditure; but the expenses attending agriculture are so much varied by contingencies, which the greatest skill and industry cannot command, that they will not yield to general computation. The unfortunate death of live stock has brought many a thriving husbandman to ruin: even the pricking of a horse's foot sometimes disappoints his pro-jects, and greatly inflames his expense. While a propi-tious

CLYDESDALE.]   G   tious

tious season forwards his labour, and diminishes his outlay, an unexpected turn of bad weather overturns what
he has been doing; his labour is lost, and he must go
over the same ground again, perhaps with less prospect
of advantage.

As the expense of cultivation cannot well be calculated, so neither can the profit. We can neither calcu-
late how much corn the frost will blight, the wind will
shake, or the rain will rot. We may safely assert, however, as has been already hinted, that the profit is not,
in general, adequate to the stock and industry employed,
or the toil and hardship undergone in quest of it. Such
comforts as farmers enjoy, and such savings as they have
made, seem to arise chiefly from an unremitting parsi-
mony, from a minute attention to the detail of their
business, and a remarkable knack of making the most of
every thing in which their interest is concerned; habits,
which the high rents and accompanying difficulties have,
no doubt, taught and confirmed. Fifty years ago, the
farmers of this county lived, as a celebrated poet ex-
presses it,

> " Like the gay birds that sung them to repose,
> " Content, and careless of to-morrow's fare."

Their rents were low, and so was their ambition; and
they jogged on in humble ease. But when the rising
prosperity of the nation began to hold out superior pro-
spects of advantage, they were, on a sudden, seized with
a rage, which could scarcely have been expected to be
found among them. They eagerly outbade one another
for every farm, as leases fell, and raised, at the same
time, the rents and the avidity of landholders to a great
pitch. Such of the new lessees as were unequal to the
task, soon sank under it; but many have struggled
through, and a body of men is now formed, very well
adapted

adapted to their laborious employment. Such has been
the advantage of the rapid rise of rents; but this has
been much overbalanced by the effect it has had, first,
in diminishing the capital employed in agriculture, and
next, in retarding its natural and proper increase. Nevertheless, there is reason to hope that the present race of
husbandmen, with a very moderate degree of countenance and encouragement, will gradually arrive at a situation fitted to bring the country to the greatest possible
improvement. Of late, however, the rent of land has
risen to such an extravagant pitch, that it is to be feared farmers will not be able to hold their ground; and
thus the progress of agricultural improvement will receive
a fatal check.

CHAP. V.

ENCLOSING AND DRAINING, ARABLE AND GRASS GROUNDS, IMPLEMENTS, &c.

IT is the universal opinion here, that the easiest
and surest way of increasing the fertility of land, is to
let it remain for a considerable time in pasture, so soon
as it has been put in condition to bear abundance of
grass; and that the richer it has been made by manure,
when it was laid in grass, its fertility will increase the
faster. At the same time, it is found, that land is ren-

dered

dered more productive, by taking it, at intervals, from pasture to tillage, by which the vegetable substances accumulated on the surface, are incorporated with, and enrich the soil, tending also to open and separate its parts, when too dense and tenacious, and to give it additional mucilage to retain the moisture, when too open and dry. Hence it is, that alternations of tillage and grass are now the general practice throughout the county, little land being kept long in a course of tillage, without being laid in grass, and none that is thought fit for tillage left in perpetual grass. The little swampy plains, among the eminences on both sides of the county, are the only instances of perpetual meadow.

For this reason, the articles in the title of this Chapter, which, according to the plan given by the Board of Agriculture, should be the subject of four chapters, is comprehended in one. Enclosing is the basis of the present economy through the most of the county; draining and supporting of fences are parts of the husbandman's employment; alternate tillage and grass are the chief objects of his attention; and the implements of husbandry are the instruments with which his work is performed. In treating these subjects, most of the articles comprehended in Chapter XII. of the arrangement proposed by the Board, will be brought in, and make that head less necessary in the present work.

According to this plan, the present Chapter will be divided into the following general Sections, viz.

1st, Enclosing and draining.

2d, Tillage, or cultivation, with the implements of husbandry used.

3d, Manures, or substances applied to fertilize the soil.

4th, Plants cultivated, and rotation of crops.

5th,

5*th*, Grass grounds of all kinds, with the purposes to which they are applied.

Each of these will be branched out into as many articles as may be requisite for the illustration.

SECT. I.—ENCLOSING AND DRAINING.

ART. 1. *Enclosing.*—Enclosing is recommended and enforced by several old Scottish statutes, which, as they do not appear to have taken great effect, it will be needless to quote. It is appointed by statute in 1661, cap. 41, that adjacent proprietors shall be at equal expense, in making fences, with the proprietor enclosing his property along the march between them; and by statute the same year, cap. 17, and 1685, cap. 39, judges and magistrates are authorized to straight marches between conterminous properties, to make enclosing less inconvenient and expensive. These laws have been frequently resorted to, and greatly forwarded enclosing.

The advantages arising from enclosing, seem not to have escaped the observation of the inhabitants of this county at an early period. The remains of mounds, probably made to divide the land kept in culture from that on which the cattle pastured, may still be traced in different parts. Even where they no longer exist, the memory of them is still preserved, through all the country, in the names of many places compounded of the word *dike*. How these dikes or fences were constructed, how far they answered the purpose, or why they were abandoned, is no longer known. It is certain the spontaneous growth of shrubs, which would tend to make them more defensible, is now ceased in places where it once prevailed.

G 3

prevailed. A few stone fences, of an old date, here and there still remain. On the low grounds where shrubs thrive, there are a good many old enclosures, fenced with hedges composed of various kinds; and some of a considerable standing, fenced with the white thorn alone.

The spirit for enclosing, which seems for a long time to have been in a great measure suspended, revived about 50 years ago, and has increased, and proceeded with great perseverance ever since. There is scarcely any place where the land has been deemed improvable, and capable of bearing hedges, but some attempts towards enclosing have been made. The most common mode of enclosing is with ditches, pretty generally known by the name of clap ditches, having a row of white thorn plants laid in the face of the mound formed of the earth taken out of the ditch. Though one has continued to follow another in this practice, it has proved, on the whole, very unsuccessful. In fertile soils, the roots of weeds protected and fostered under the mound, perpetually put forth their shoots, and injure the young thorns. In clay soils, the argillaceous substance at the bottom, which is the bane of many kinds of plants, and of the white thorn as much as any, surrounding the roots of the young hedge, on all sides, as soon as they reach it, checks the progress of the plants; at the same time, the mound of dense earth excludes the influence of the sun and rain; and the hedge, which promised to grow at the first, becomes stunted and puny in a few years. Hence, except in a few places where the soil, bottom, and exposure, are uncommonly favourable, there are few hedges in the county defensible, without perpetual and expensive repairs with dead wood. The expense of making these ditches, with plants and a cocking of

wood,

wood, is from 2*s.* to 2*s.* 6*d.* per fall, a measure of 18 feet 6 inches, used in the country. When it is considered to what extent enclosing has been carried, it will appear, there has been a great deal of money very unprofitably laid out. Some people now begin to be sensible of the general error, and, instead of clap ditches, make mounds solely of earth collected from the surface, faced up with stone or green sod, mix the collected soil with manure, and plant the hedge on a border along the top. This seems to be the manner in which all the old hedges, that remained good, have been done. As the white thorn, though it makes a formidable hedge, when it thrives, makes but a poor figure in a barren soil and exposed situation, it is now common to plant a third or fourth part of beeches, intermixed with the thorns; the former being found to be a hardier, more thriving plant, than the latter, and consequently better adapted to shelter a stormy country, as well as to strengthen the fence. It is probable this may be a considerable improvement in enclosing; but there are often so many barren spots, from the out-skirting of the mineral strata, and the like, occur in the lines where it would be wished to draw fences, on which no plants can thrive, that enclosing can never be so general, or so sufficient as it ought to be, till stone walls be more in use. It is unfortunate that much of the soft stone found near the surface moulders in the air, through time, and is therefore less durable in fences. In the light soils of the upper ward, thorn hedges frequently fail for want of moisture. Here, raised mounds would make the matter worse. In such cases, the best way of raising thorn hedges would probably be, to summer-fallow the line on which the hedge was to be planted, about 10 feet wide, freeing it completely of the roots of weeds, and working a good quantity of dung into it.

G 4 The

The edges of this might then be turned up towards the middle, in order to thicken the soil, and the hedge planted in the centre. Elm plants, which delight in a dry open soil, and submit very well to be dressed and pruned as a hedge, might be substituted in place of beeches. Hedges would perhaps succeed in this way, where they fail in clap dikes; but there are stripes running through this kind of soil, as unfavourable as the out-skirting of the mineral strata. These are of dry sand and gravel for a considerable depth: they are called scalds by some of the English husbandmen. For such places there is perhaps no remedy but a dead fence.

ART. 2. *Draining.*—There is nothing of great consequence to be observed with respect to draining. In all the clay country, the great business is to carry off the surface water, which can only be done in open drains. The numerous ditches made for enclosing, already mentioned, though frequently not answering the intended purpose, are very useful conductors of water. The draining of clay ground is principally performed by the manner of laying out the ridges, which will be taken notice of under the following branch. Where larger receivers than the ordinary furrows are necessary, it is thought the most eligible way to make the sides of them very much shelving, as recommended by the late Lord KAIMES; but contrary to what he advises of doing them with the plough, it has been found that they are always cheapest and best executed with the spade. When land is drained, which is wet from other causes, such means as have been used in other parts of Britain, and will probably be described in other Reports, have been used here. It may only be observed, with respect to hollow or covered drains, that those which have as much

declivity

declivity as circumstances will admit, filled with plenty
of stones, and the uppermost made very small, continue
longest serviceable; so that those which are executed at
the greatest expense at first, frequently turn out cheapest
in the end. The smallest experience must convince
every one, of the necessity of freeing land from stagnant
water, so that every husbandman finds himself obliged
to pay some attention to draining. But it is a subject of
regret, that this first and very important branch of rural
industry, is seldom so much studied or so attentively
executed as it deserves. Water, which is an agent in-
dispensably necessary in the economy of vegetation, is
also, when retained in excess, the bane of all land vegetables. To support them, it ought to be in the state of a
light vapoury fluid diffused through the soil. It has of
late been discovered to be composed of two principles,
both of which indeed are constituent parts of vegetables.
But it has been proved, that the organs of vegetables can
only digest a small quantity of water. Whenever it exceeds what may serve for this purpose, and for supplying
the perspiration habitual to the different plants, it is injurious. Its effects on the soil, in winter, when vegetation is suspended, are likewise highly prejudicial, as it
counteracts the purpose of cultivation, by deranging the
texture which has thereby been given, and excludes the
beneficial influence of the atmosphere. Hence it is, that
all soils which have been soaked in standing water, in
winter, suffer most in the summer's drought. In dense
soils, covering an impervious bottom, this injury is best
obviated by giving the ridges a regular convex form,
that the water which falls in rain may flow quickly away
without soaking into the soil, and keeping the furrows
or drains between the ridges clear to receive it. In all
porous or soft yielding subsoils, through which water
oozes,

oozes, and approaches the surface, recourse must be had
to hollow drains.

———

SECT. II.——TILLAGE OR CULTIVATION, WITH THE IMPLEMENTS OF HUSBANDRY USED.

ART. 1. *Tillage.*—Summer-fallowing is practised for
different purposes. It is either with an intention to free
the ground from weeds, to give the ridges a proper form
and direction for throwing off the surface water, or to
open and mellow a dense strong soil*. In all light soils,
weeds multiply quickly, and frequent recourse must be
had to summer-fallowing to destroy them: but since the
turnip husbandry has been introduced into the upper
part of the county, the land is fallowed in the Spring
and beginning of Summer, and turnips sown upon it, in
drills. The cleaning of the ground is completed by

---

* As the question about the advantage or disadvantage of repeated
summer-fallowing has been much agitated among agriculturists, it may
be proper to observe here, that from the evidence of frequently repeated experience which cannot be resisted, summer-fallowing injures rather
than improves the texture of hard thin soils: for though the reiterated
labour in a dry season, and the fermentation of the manure, render the
soil more mellow, and of course more fertile for a few years, it becomes
in time more hard and impervious to the roots of vegetables than before; and the oftener the fallowing has been repeated, the soil is the
more injured. The perfection of the soils consists in a proper medium
between rigid cohesion and excessive friability. The proportion in
which the siliceous, aluminous, and calcareous earths, are mixed in the
soil, constitute its consistence. But there are substances lodged in soils,
which occasion a repulsion among the particles, of which the pure
earths, of themselves, are incapable. This tendency to repulsion, within due bounds, always promotes a vigorous vegetation. But the effect
of summer-fallowing is to counteract this repulsive power.

hoeing

hoeing the intervals; and an entire summer-fallow is seldomer used. In the light soils, in the lower part of the
county, the land is too high rented for the turnip husbandry; and potatoes, for which there is great demand
in that populous district, are substituted in the place of
turnips, and the ground cleaned by hoeing them attentively. In the clay soils, a great part of the land has
been summer-fallowed, principally for the purpose of
draining it, by giving a proper form and direction to
the ridges. Various forms and sizes of ridges have been
tried for this purpose; but that which seems now to
prevail most, is ridges from 13 to 15 feet wide, properly
rounded, and not raised very high. To keep clay ground
dry, it is found necessary to consult Nature, and make
the direction of the ridges follow the course of the declivity, making drains across, whenever the water does
not follow the furrow; but when water can be led away
without such cross drains, it is much to be preferred.
In the more elevated parts of the county, where the exposure is deemed too severe for wheat, the land lies
longer in pasture, and summer-fallow is less frequently
repeated. On the lower grounds, husbandmen go round
their farms, summer-fallowing as much yearly, as they
can procure manure for, in order to sow wheat. Part of
the light land, both in the upper and lower parts of the
county, undergoes a spring-fallow for barley; but barley has succeeded so ill, for many years past, on the
clay grounds, that the culture of it is almost abandoned.
This perhaps is much owing to the farm dung being
mostly consumed for raising wheat. For all other crops,
seldom more than one ploughing is given. In some
parts of the county, the ploughing is begun soon after
the harvest is over, but it is more general not to begin
till after the 1st of January; and in frosty or very rainy
seasons,

seasons, there is frequently much land to plough when March comes in.

In heavy soils, it is common to put three or four horses to the plough. Some husbandmen have made a late improvement, of making the fore and hinder horses draw from different points of the beam, so that the two lines of draught may coincide. In lands that are light and easy, in the latter ploughings of summer-fallow, or in breaking up turf, with an ebb furrow, two horses without a driver are now frequently used. There are, however, many intelligent husbandmen, occupying the heavy clay soils, who make very little use of the two horse plough: nor do they reject it from prejudice, but support their conduct by cogent arguments, which it may be proper here to state, and leave them to the judgment of the reader.

The clay soils in this county have almost always a dense argillaceous substratum (generally called *till*), so much of the same nature with the soil above it, that the particles of the latter, which are washed down with the rain upon the former, assimilate with it into one mass, and the till bottom seems to approach to the surface. When this is allowed to take place, rain, when it descends, is retained upon the top, the roots of plants are chilled by its stagnating there, and the crop fails; for which reason, it is found necessary to plough the ground, once or oftener in every rotation, very deep, in order to allow the water to descend through the opened ground, and the roots of plants to expand freely. But though the improvements made in the construction of ploughs has considerably increased the powers of draught, so strong is the adhesion in these clay soils, that a pair of the best horses are frequently unable to overcome the resistance, when the ploughman aims at the necessary depth.

depth. He is naturally led, therefore, to lighten the draught, that the horses may more easily proceed; and the work comes insensibly to be more lightly executed than circumstances require. It is, therefore, thought necessary to have four good horses in the plough, with a boy to drive them, that the ploughman, having nothing to attend to but the execution of the work, may make it of a proper depth and regularity. The husbandmen alluded to, farther urge, that the advantage of two horses in a plough without a driver, instead of four with one, is rather apparent than real, in heavy soils where the resistance is considerable; for they assert, that, when the days begin to lengthen, and the ground becomes dry, four horses and two men will plough an acre and an half in a day, and that one man, with a pair of horses, will not execute more than the half in the same time. They do not, however, altogether reject the two horse plough, using it generally for the latter ploughings of summer and spring fallow.

Even where the soil is less dense and obdurate, it would probably be proper to plough somewhat deeper than ordinary, every third or fourth year. In all soils, the finest parts must be washed to the bottom of the stirred ground, and the use of it be lost while it remains there. If a ploughing deep enough to bring this up cannot be managed with two horses, more power should be added. A pair of oxen might be kept on a farm for such purposes, to great advantage, and very little additional expense.

ART. 2. *Implements of Husbandry.*—The ploughs used here are, 1*st*, The Scotch plough: this well-known instrument has long been esteemed the best for ploughing stiff or stony land. It was preferred to some of the later invented ploughs, as setting up the furrow with a bold shoulder, and so furnishing a plentiful mould to

cover

cover the seed. Some improvements have lately been made on it, particularly the head, beam, and stilts or handles, are made shorter, in order to make it easier drawn. It is still found, however, to require greater power to work it than some others, and therefore, since ploughing with only two horses is become more prevalent, this plough is in less esteem.

2*dly*, A little plough, brought to this county many years ago from Northumberland : it seems to be the same with that described by Lord KAIMES, under the name of the Rotherham plough, and has been found to answer very well for stirring of fallow.

3*dly*, The Rutherglen plough, invented for the purpose of turning up the deep soil of the valleys with a strong furrow : it has been used chiefly in the neighbourhood of that town.

4*thly*, SMALL's plough : the improvements which this ingenious mechanic made in the construction of the plough, made it easier drawn, and it quickly obtained reputation, and was generally adopted. But it fell short of the perfection aimed at, and has undergone many modulations, almost every plough-wright having his own particular cast of the mould-plate, &c.

5*thly*, WILKIE of Uddingston's plough : this profound artist has had the good fortune to discover every thing requisite to the construction of a complete plough, and formed an implement the best adapted to the purpose, of any that has yet appeared in this county. It turns up the furrow with a bold shoulder, like that for which the Scotch plough was valued. Its shape, like that of a well-tapered wedge, turns over the soil with the application of very moderate powers; and its broad-winged share leaves nothing unstirred. There are instances of other different con-

constructions of ploughs, but they seem only to be variations of the kinds above-enumerated.

The common harrows, which are still the most generally used, with four bills or beams, containing 20 teeth,
are much the same as have been described in the Reports
from other counties. It is found necessary, both for re-
ducing stiff ground, and collecting the roots of weeds, to
give the teeth a considerable bevil forward, so as to stand
at an angle of from 70° to 75° with the plain of the harrow. There are others heavier, commonly called *brakes*,
of different weights and constructions, according to the
fancy of the owner, and the purposes for which they are
intended. Of late, pairs of jointed harrows have been
introduced, each having three bills, and the pair connected by joints, by which, while they are kept together,
they are allowed to ply to the surface; the teeth are also
placed so as not to follow one another directly in the line
of draught. These are drawn by a pair of horses, and
have been found to be very executive.

The roller is an important implement in the culture of
the fields. Besides smoothing the surface, and bruising
clods, to forward pulverization, the use of it can never be
too much recommended for condensing open soils, in the
droughts which frequently succeed the seed time. Even
in the heavy soils, which are for the most part but too
solid, the application of the roller is of great importance,
during the droughts of the spring. By pressing down
the mellowed clods around the roots of young grass and
wheat, the plants are re-animated, and a fresh luxuriancy
soon appears. By compressing the surface of fields sown
with spring corn, which, however solid they may naturally be, heave with the spring drought, the interstices
are closed, the moisture retained, the roots of the corn
fastened, and the progress of the vermin which prey upon
them

them checked. Mr. Cook's discovery, published some years ago, that snails and slugs, which come to the surface after the sun goes down, may be destroyed by rolling in the night, merits attention. The rollers here are of free stone, or of solid timber, and very rarely of cast iron, this last being too dear for common husbandmen; but the most approved rollers, and which are now getting pretty much into general use, are hollow cylinders built of wood, the circumference clothed round with strong plank; the diameter is about three feet, and the roller is divided into two equal parts, which turn round on an iron axis. The largeness of the diameter makes the draught so easy, that one horse can pull as much weight as two could do of solid stone, and the division of the roller into two parts facilitates the turning, the half on the inside moving back, while that on the outside comes forward.

The drilling instruments are, the turnip drill, and one which, by changing a nut upon the axis which turns round at the bottom of a hopper, sows either beans or smaller grain. Both of these sow only one drill at a time, and are used chiefly in the upper part of the county, the heavy soils lower down being less adapted to the drill husbandry. The instruments used for horse-hoeing, are small ploughs of different constructions, all of them very simple.

It is needless to describe the spade, the hand-hoe, the wheel-barrow, &c. simple instruments which, in the hands of the dextrous and intelligent labourer, are perhaps not much less important than all the machinery which has yet been invented for cultivating the ground.

An instrument composed of two sticks joined by a pin, and resembling the smith's tongs in appearance and use, is applied to pull thistles and docks in the corn fields; but the dock-iron is the fittest instrument for clearing

grass

grass grounds of docks. In all cases where the hoe cannot be used, if smaller weeds appeared to prevail so much as to injure the crop, women and children used to be employed to pull them with the hand; but from the scarcity of such hands, and the high price of labour, this is almost given up. Husbandmen now, when annual weeds appear to prevail much among the spring corn, harrow the ground while the weeds are young, and their roots have not taken a deep hold. In this way the most of them are destroyed, and the corn which is deeper rooted, so far from being injured, is benefited; for though a few plants be torn up, the rest are invigorated by the stirring of the earth, and thrive and tiller more abundantly. The same practice is successfully used with pease and beans. If harrowing has been neglected, and wild mustard, which is the most frequent weed, prevails, the flowery heads are cut off with the scythe, when it is in full blow, without injuring the corn.

The carts of this county are still mostly of a plain and simple construction, at the same time light and strong; but of late, many alterations and new constructions have been introduced, which probably tend much more to increase the expense than to add to the conveniency. Indeed, refinements of this kind seem to be unwisely followed in the constructions of many implements of husbandry, and particularly in carts, harnessing, &c. so that a horse seems to groan under a load of iron and leather, without any good purpose being served. But the iron axle ought certainly to be excepted, as it compensates in durability, and diminution of friction, the expense of purchase and the weight it adds to the carriage. The general effect, however, of all such refinements, is to abstract a part of the stock of husbandmen from its proper employment. Carts are drawn by a single horse, experience having

CLYDESDALE.]   H   evinced

evinced that, in this way, the animal is capable of the greatest exertion.

The sickle is almost the only instrument used in reaping. Several mowing instruments have been introduced, but soon given up; and now that thrashing-mills are coming much into use, it is probable the use of the sickle will be still more confirmed, as corn thus reaped is in best order for thrashing in the mill.

Every farm has fanners, and there are now a good many thrashing-mills in different parts of the county. Where these are wanting, a considerable part of the corn is thrashed with flails by the farm servants, in the winter mornings, by candle light.

SECT. III.—MANURES, OR SUBSTANCES APPLIED TO FERTILIZE THE SOIL.

THIS section may be divided into three articles, treating, 1*st*, Of the manures used; 2*dly*, Of their effects; 3*dly*, Of substances which may be used as manure.

ART. 1. *Manures used.*—Little marl of a valuable quality has hitherto been discovered in this county; some has been found under the mosses in the elevated parts of Carnwath parish and Lesmahagow, and laid upon the land with good effect; but the land is too high there to encourage the culture of corn to any great extent, and the distance too great, to carry the marl with advantage to the corn lands lower down. Marl of an inferior quality is also seen in several places in the lower part of the county. It lies most commonly between two stratas of the free-stone rock, and would probably be found expensive

pensive to work. Besides, it is only found in those parts of the county where clay is the prevailing soil, and these clay soils show symptoms of possessing calcareous substance, by effervescing with acids; so that clay being also the predominant part in the marl, there is probably too strong an affinity between the manure and the soil, on which it could be most conveniently applied, to produce any considerable effect. Lime, therefore, is almost the only fossil used for manure, and it is now become very dear, as much as will load a single-horse cart being sold at the kiln from 6s. to 8s. It is applied either upon fallow or grass grounds, at the rate of from 300 to 600 Winchester bushels per acre. The first time land is limed, its fertility is visibly increased. If it is moderately cropped, and allowed to rest for several years, the effects of the second liming are still more considerable; but all after limings have very little effect, and there is now land in this county, on which it proves quite useless to lay lime alone. For which reason, those who cannot procure enough of other manure, compound lime with scourings of ditches, cleanings of roads, and some kind of surface earth having a close turf of grass, with a little dung between the layers of earth*. This has been found to answer the expense, when lime alone would not. It has been found very beneficial to lay lime upon well swarded pasture, and allow it to lie on the surface for two or more years before the land be ploughed and cropped. Where few corn crops were taken at a time, and the land left long in grass, before being again broken up, the meliorating effects of this practice have been almost incredible. Besides lime, and the dung, and compost made about the

---

* It is studied, as far as circumstances will admit, to lay compost of the lightest quality on the heaviest soil, and *vice versa*.

farm,

farm, horn-shavings, soot, woollen rags, parings of leather, peat, the ashes made by burning it, and coal ashes, when they can be had, are used as manures; and in the neighbourhood of a town, dung is brought from thence. From the city of Glasgow, in particular, dung is carried for six or seven miles round.

Some experiments of gypsum, for a manure, have been made; but the expectations raised by the accounts from America, Germany, &c. of the great success attending this practice, have been disappointed, the application of gypsum having produced little or no additional fertility.

Irrigation, by streams of water led over the ground, has been little practised here, except on the little swampy meadows above noticed; and these are flooded rather with a view to the substances which the water carries along with it, and deposits on the meadow as it glides along, than on account of the fertility which water itself bestows. It is not probable that irrigation will soon become common in this county. In the upper parts, where the ground is light and the subsoil open, it might, no doubt, be advantageous, wherever a command of water could be had; but in the clay soils, where the ridges are rounded for the sake of surface draining, it would be difficult to spread water over the surface; and if it could be done, the sweet herbage would soon be destroyed, and coarse aquatic plants raised in its place. The general practice of alternate tillage and pasture, is also inconsistent with watering. Irrigated grounds should be kept in perpetual meadow. In other counties of Scotland, where grass grounds have been irrigated, with a view to tillage crops, a few abundant crops have generally been produced, but the ground left in a wretched state after them.

ART. 2. *Of the Effects of the Manures used.*—The old husbandmen did not spread the lime upon the ground till

it

it was completely soaked with water, and stuck together in lumps. It is probable they were frequently in the right. The poor soils, on which it was often laid, possessed very little vegetable substance on which caustic lime could act; and the principal effect of the lime likely was, to add to the calcareous substance when it was deficient, and to separate the parts of a soil which was too dense; by which the land was better prepared to receive advantage from dung, and such like manures. Accordingly we find, that lime alone, when laid upon poor soils which produce little herbage, has but a slender effect in increasing the fertility. It is agreed, however, that though the stalk is not much more vigorous, the grain is plumper, and the herbage is sweeter, when the land is laid in grass. Lime is now more frequently laid on while it is in some degree powdery; and when it is applied to grass grounds, the surface of which are covered closely with a growth of mosses and decayed herbage, the effects of it very soon appear, by consuming the old turf, and raising a pleasant verdure. But to derive benefit from lime on poor land, it is found necessary to compound it with good surface earth, or mossy earth, and in this way it has never failed to succeed. Lime is also laid on fallow ground, but the effects of dung are always more powerful. That very ingenious and industrious nobleman, the Earl of DUNDONALD, to whom the public is indebted for so many valuable discoveries, spent the summer 1794 in this county; and, with his usual zeal and activity for the public good, made a number of experiments on substances which might be used as manures for different soils; a treatise on which he soon after published, and in this the inquisitive reader will find a great deal of information. His Lordship found that peat earth, hitherto thought so useless, might be rendered valuable

manure,

manure, by mixing it with new slaked lime; and this is now begun to be put in practice.

Not that peaty substance had not been applied as manure before that period. It had long been the practice to mix such substances with lime or dung, or both, in making composts, and sometimes to spread it on the surface by itself. But the rotten substance about the verge of mosses, upon which the water from the higher grounds had carried mud, and raised a thick growth of herbage, was chiefly used. His Lordship recommended the pure quick peat, which had hitherto been thought unfit for the purpose.

Since this Report was last published, Lord MEADOW-BANK, having made great improvement on a thin hard soil, in a neighbouring county, by the application of peat, fermented with new farm dung, has generously published his process of preparing that substance: and his example is now successfully followed by different husbandmen in this county. His Lordship's general rule is, to mix three or four cart-loads of peat, pretty accurately, with one cart-load of the dung, to cut and divide the peat, and lay the whole up loosely, so as not to check the fermentation of the dung. The compost heaps not to exceed fifteen feet wide by four feet and a half high: the fermentation of the dung ought to occasion a heat in the moss which would raise the mercury in FAHRENHEIT's thermometer to near 90°; if the heat exceeds that, the fermentation should be checked by watering, or adding more peat. After this heat is somewhat subsided, the heap should be completely turned; and in about three weeks after, a new fermentation, with more moderate heat, will ensue. The compost is then in fit condition to be spread on the ground; and it will have as good effect as equal bulk of

the

the best dung. Lord DUNDONALD's theory is, that the alkaline quality of quick-lime, renders the vegetable substance of peat soluble in water, so as to ascend in the vessels of plants, and promote their growth. But Lord MEADOWBANK asserts, that in the course of his experience, he has not found lime of any use in dissolving peat. His theory is, that the insolubility of peat proceeds partly from its being long soaked in cold water, but chiefly from the antiseptic quality of the tan and gallic acid with which he thinks it is charged. He seems to think, that the vegetable food is formed in the course of the hot and putrid fermentation, and that this food consists chiefly of secondary compounds. " Fermenta-
" tion (says he) decomposes the mass of animal and ve-
" getable substances, by the escape of part in gas, by
" imbibing somewhat from water and from the atmos-
" phere, and by the formation of compound matters
" from the re-union of the parts into new combina-
" tions."

Horn-shavings, are a powerful manure on land possessed of vegetable substances, but the effect is less considerable on very poor soils.

Woollen rags act powerfully for one year. Parings of leather tend to open heavy soils; but they rot slowly, and have not a sudden effect.

Peat ashes are either made by paring and burning the surface of moss, in order to fertilize the ground itself, or by burning the ground on the borders of mosses in heaps, and carrying the ashes to the firm ground near. In the former case, very great fertility is produced at first, but, in a few years, these ashes, instead of producing useful plants, bear nothing but a kind of tall growing moss. In the latter, fertility is increased for a few years without

any

any visible bad effects afterwards*. But pure peat earth
cannot be burnt till it is cut in small pieces, and thoroughly
dried, and it yields but a small quantity of ashes. Ashes,
in quantity, can only be made of the rotten earth about
the borders of mosses, in which there are a great deal of the
living roots of plants. Coal ashes are best when soaked
with urine, &c.

The dung collected in towns is the most powerful manure. The dung of the farm-yard, that is, the dung and
litter of the live stock, has sometimes surface earth or
moss mixed in it, to increase the quantity. It is carried
to the fields, sometimes twice, sometimes only once, in
the year, but never turned. The husbandmen of this
county are surprised to hear it recommended from other
places, to turn dung over, and lay it up loosely, to admit
the air and hasten the putrefaction; as they find that the
dung which has been most condensed by the trampling of
cattle is the best.

It would be of much importance to the husbandman, to
know what ought to be the effect of the substances which
he applies as manure. Unfortunately the minds of the
people of that profession are too much occupied in the
ordinary routine of their business, to leave them leisure
for making observations sufficiently discriminating to
correct the errors which tradition has handed down, and
the doctors and writers in agriculture have taught on the
subject. It may therefore be proper here to give a few
hints, which are founded on experiments, and justifiable,
it is believed, on the principles of science. Lime, so
frequently applied as a manure, cannot be productive of

---

* There are always many particles of a kind of charcoal in peat ashes,
which are partly soluble in water: and this seems to be the chief cause
of their fertilizing quality.

vegetable

vegetable food: for though it is the species of earth which has been found to abound most in the analysis of vegetables, yet the proportion found in any is very inconsiderable; seldom perhaps a thousandth part of the bulk of the whole plant. And for any thing we know, that small part may not be derived from the soil; since we find an efforesence of lime on the external bark of the ash, even when we can discover no traces of it in the soil where the tree grows. It is, however, an excellent ingredient, as it tends to separate the parts of a dense soil, and fill the vacuities of a porous one; and hence to improve the texture of both. Other earths may sometimes be advantageously applied for the same purpose. A large dose of sand laid on a strong clay soil, after it has been reduced by repeated ploughing and harrowing in a dry time, has a happy effect. Clay laid on blowing sands, and allowed to lie spread on the surface till it moulder by the influence of the atmosphere, is also beneficial. Marl improves the soil just as other earths do. Animal and vegetable offals, known by the general name of dung, being either directly derived from vegetables, or from animals whose food is ultimately obtained from the vegetable kingdom, yields positive food to growing plants. And it is this upon which the fertility of countries long cultivated chiefly depends. The management of this substance, therefore, deserves particular attention. For this purpose the nature of vegetables ought to be studied. All vegetables are compounds made up, 1st, Of principles, which, when freed from combination, are aeriform; 2dly, Of charcoal; 3dly, Of a small quantity of earth. This is proved by submitting any vegetable to combustion. If the fire is smothered, the aeriform principles only escape, and a bulky residue, consisting of charcoal, debased with a small proportion of earth, is left; but if the fire burn freely

in

in the open air, the charcoal, combining with air, also
escapes, and nothing remains but a small quantity of grey
ashes, which is the incombustible earthy part of the plant.
Now the charcoal of almost all vegetables come to full
maturity when the aeriform principles are expelled from
it, is incorruptible and insoluble in water, so that it can
yield no food to vegetables. Hence it appears, that the
different principles of such compounds as are to be used
for manure, ought as much as possible to be preserved,
and kept together, by moderating the fermentation which
such compounds undergo when falling into dissolution.
For fermentation being analogous to combustion, when
the external air is freely admitted into the fermenting
body, it combines therewith, a strong heat ensues, the
aeriform principles are set at liberty, and fly off accompanied with the most soluble part of the charcoal combined with air. The remaining charcoal, become rigid
by the effects of a violent fermentation, is partly insoluble, or at least very refractory; so that the residue
of the fermented body, though an exceeding fine mould,
retains but little direct vegetable food. To prevent this
waste, the substances of which manure is to be prepared
should be macerated in excavations holding putrid water,
and compressed to exclude the too free admission of the
air; at the same time all extraneous water should be excluded. This doctrine, so opposite to what has been
taught by some philosophical agriculturists, is here shortly
stated for the consideration of practical husbandmen,
who, it is believed, will always have found the most
powerful manure towards the bottom of a wet dungstead.
Horn-shavings, urine, and soot, besides holding charcoal
in solution, contain a principle capable of dissolving the
charcoal of other bodies. When they are to be applied
to a soil containing little vegetable food, they should be
                                                      mixed

mixed with other substances. Part of the charcoal of soot is difficult of solution. To assist its solution, it should be agitated in water and poured on composts. Ashes, when they are not accompanied with soluble charcoal, yield no direct vegetable food, but are an excellent ingredient in the soil. Their chief use, however, is as a vehicle for the conveyance of putrid liquids, which they eagerly absorb. Composts are not only useful as they contain the aliment of vegetables, but also as they add to the thickness of thin soils, and correct the texture and consistence of such as are defective in that respect, especially when the substance of the compost is of an opposite consistence to that of the soil. But, perhaps, peat is of all other, the best ingredient of compost for hard cohesive soils, when it can be conveniently had. Being a vegetable substance, its more soluble parts go directly to the support of growing plants, while its refractory fibres separate the particles of cohesive earth, and by communicating a black colour, enable the soil to absorb the rays of the sun, and receive a greater degree of heat. But the great degree of incorruptibility which peat possesses, has been a common objection. We have directions how to overcome that, from two great authorities lately cited; different indeed, but perhaps both right. Experience has proved that six parts, by bulk, of peat mixed with one of new slacked lime, and repeatedly turned, and intimately mixed together, before the lime become effet, turns to a fine mellow friable mould, which, when applied to the soil, greatly promotes vegetation. Nor is there any doubt of the fertilizing effect of peat fermented with dung. Those two composts do not differ more from each other in value, than as the value of the calcareous earth in the one differs from that of the direct vegetable food contained in the dung of the other. But Lord MEADOW-

MEADOWBANK does not seem to be correct, in supposing
that the secondary compounds, formed in the fermentation of the dung mixed with peat, is the chief food which
vegetables derive from the compost. It has been proved
by the experiments of some illustrious philosophers, that
secondary compounds, or salts, are the poison of plants.
If those salts were in sufficient quantity, they might have
some effect in resolving the peat into vegetable food.
Indeed, whatsoever alters its original conformation, and
destroys its habit of holding fluid water like a sponge,
does the same. Peat which has been full exposed to all
the vicissitudes of the weather through the winter, when
put in a flower pot, corn planted on it, and duly watered,
will produce healthy plants and ripe grain ; but it shows
much greater fertility when mixed with solid earth.
Neither is tan and gallic acid the cause of the sterility of
peat. A plant of mint, placed on a vessel filled with the
juice squeezed from a bed of peat two feet under the surface, continued to put forth new leaves ; and these leaves
were broader, and of a deeper verdure, than those of another plant of the same size placed upon plain water.
The porous conformation, the excess of fluid water it
holds, and the cold thus produced, are the causes of the
sterility of this substance ; and it is by the heat and gase-
ous principles which escape from the fermenting dung
penetrating into the interstices of the peat, and altering
its conformation, that its sterility is reclaimed by Lord
MEADOWBANK's mode of preparation.

ART. 3. *Of Substances which may be used as Manures.*
—It has been already observed, that animal and vegetable
offals are the genuine food of growing plants. The
quantity of such offals scattered about as nuisances to
shock the delicacy of the passengers, show an equal disregard to cleanliness and economy. Were all the spoils
of

of both kingdoms, liquid as well as solid, which are thrown off as useless, carefully collected and carried to the fields, what an increase of fertility might be expected! But besides these, there is no situation in which means are wanting to make considerable augmentation to the ordinary quantity of manure. Rushes, fern, the musci which are frequently found in woods and plantations, steep banks and waste grounds, might be gathered, dried, and laid up to absorb the urine of cattle and other putrid liquids. All kinds of weeds and inesculent herbage might be collected about the time of flowering, before the fibres of the stem become rigid and refractory. The young and tender shoots of trees and shrubs, not ripened into perfect wood, the leaves of trees shed in autumn, the grassy bottoms of open drains, pond and river weed, and sea weed near the shore: all these might be laid in regular strata, along with the farm dung, in a wet dungstead. There is some cause to believe, that the earth frequently contains in its bowels substances capable of fertilizing its surface, if the proper application of them were understood; and it would be a matter of great importance, that experiments were set on foot for making such discoveries. The blaise or shiver, accompanying the coal, mentioned in a former part of this Report,—a kind of indurated schistus, which sometimes skirts out in the faces of banks, and moulders down with the weather. At first it appears quite barren, but after lying some time exposed to the influence of the sun and air, it not only fertilizes the soil on which it falls, but, where it accumulates to a body, shows great signs of fertility itself, plants growing in it with great luxuriancy. The noble Lord above mentioned *, strewed one kind of blaise which he

---

* Lord Dundonald.

found

found here, reduced to powder, about the roots of growing corn, and in a short time the deep verdure and luxuriant growth distinguished that part from the rest of the
field.   These schists might be had at coal mines, and at
many places where they skirt out at the surface; and
when they were found, upon fair trials, to have a fertilizing effect, might probably be used as top-dressings with
great advantage.   The same noble Author, and all the
writers on chymistry, agree, that magnesia promotes, in
a very considerable degree, the growth of plants*.   Much
of it is said to be contained in the steatities, or soap rock,
which is found in different places of this county.

SECT. IV.——PLANTS CULTIVATED, AND ROTATIONS OF CROPS.

ART. 1. *Plants cultivated.*——Wheat is cultivated on all
parts, by intervals, which are thought favourable for that
grain.   On open porous soils, sweating bottoms, and
elevated situations, the plant has not been found to thrive,
nor the ear to arrive at full perfection.   It is sown either
on fallow, or after potatoes, and more seldom now, after
oats, or pease and beans.   The time of sowing is from
the end of August to the first of November: spring
wheat is seldom sown, and seldomer succeeds.

---

* It has lately been asserted by a very great authority, Dr. TENNANT
of Cambridge, that the mixture of magnesia in the soil destroys, instead
of promoting, vegetation.   This, however, Mr. HEADRICK opposes, upon
seeming good grounds.   From some experiments made by the Writer of
this Report some years ago, he is of Mr. HEADRICK's opinion; but he
has others now depending, by which he expects to ascertain the matter.

Different

Different kinds of seed wheat have been used, particularly the bearded wheat, which prevailed much for
some time; but there is now scarce any other than the
common white and red, the seed of which farmers mutually exchange with one another. The fallowed ground
is manured for the wheat crop with dung from towns,
when within reach, or with farm-dung and lime, sometimes with horn-shavings (a quantity of which is annually commissioned from Ireland by the husbandmen of
the county), and less frequently with woollen rags.
Wheat seed is commonly steeped in a strong pickle of
sea-salt and water, and dried with hot lime, sometimes
sprinkled with urine, or the mephitic water of a dunghill, and dried the same way. This is intended to prevent the smut; but a number of experiments concur in
proving, that the caustic lime incrusted round the seed,
has the principal effect. It ought here to be observed,
that wheat sown after the middle of October, is always in
more danger of suffering from the smut, whatever pre-
cautions may have been taken, than that which has been
earlier sown. The quantity sown is from seven to twelve
pecks, Linlithgow measure, per Scots acre; the produce
from eight to sixteen bolls of the same measure.

Oats is the principal Spring corn; from two-thirds to
three-fourths of the land tilled is sown with this seed.
Different varieties of this grain are sown. Two have
been known in the country for time immemorial; namely, late seed, sown in the lower and earlier grounds:
this seems to be much the same with what is called Hal-
kerston or Angus oats in the neighbouring counties;
the other is called early seed, and is sown on higher and
later grounds: it is an inferior grain, but ripens quicker,
and produces a good deal of straw. To these may be
added, early Tweedale and Blainsly oats, both of which
have

have been long sown here. Their time of ripening is between that of the two former, and they are not deficient in
straw. There are several other kinds of early oats which
have been more recently introduced; such as the Polish
and Essex oats, the Friesland or great Dutch oats, and
the red oats; this last has indeed been long known about
the head of the county, under the name of small barley
corn. To these may be added a kind of oats called the Po-
tatoe oats, introduced into the county since this Report was
last published. It is a fine plump grain; and when sown
on a rich tender soil, the culture has been uncommonly
successful. These ripen early, but the straw is short,
and unpalatable to cattle as fodder. Some of them
shake very easily with the wind; but they are still of
importance in backward seasons, as they do not fail to
ripen and produce plump grain. Oat seed is never
steeped; but some have spread it on the barn floor, and
sprinkled it with salt, turning it from time to time, till
the salt had liquified and moistened the grain. This
practice should be more general, as it certainly has a tendency to defend the young corn from the devastations
of the numerous tribes of the caterpillar kind, which
lodge in the earth in the Spring time. A very intelligent correspondent, whose accurate observation and
veracity may be depended on, informs, that he has long
been in the custom of bringing seed oats from the sea
coast, where the air is somewhat impregnated with the
saline particles raised in spray, and that he has uniformly
observed the crops of this seed escaped the injuries of
land vermin, while others around suffered greatly. If
the small quantity of salt this grain could draw from the
air, while growing, is thus effectual, it may be presumed
that soaking the seed with salt should not be less so. It
is certain, that salt destroys all the kinds of vermin
which

which creep in the earth, and devour the roots of plants in the inland country; and it is a pity that, for the sake of a tax so unproductive of revenue, husbandmen should be deprived of the benefit of it. From 12 to 18 pecks of oats, of the ordinary measure of the county, which is somewhat larger than the Linlithgow barley measure, is sown on an acre; and the produce varies from 4 to 18 bolls per acre. Pease and beans seldom come to perfection on high exposures, and therefore are chiefly cultivated on the lower grounds. They are sometimes sown separately, sometimes mixed, and very rarely in drills. The beans sown here are ordinary horse beans, and a kind of late grey pease usually accompany them. Husbandmen frequently bring a change of seed from the Kerse ground on the banks of the Forth, which is found to be an advantage. Grey Hasting pease are sometimes sown alone; the straw of these is not so bulky, nor the fodder so good, as the late. Beans are sown as early as the season will permit, sometimes on the surface, and ploughed in with a light furrow; from 14 to 18 pecks, wheat measure, are sown on an acre. They are sometimes very productive, yielding 18 bolls of the same measure an acre: in rainy seasons, or when autumnal frosts come in early, they are sometimes good for little.

Small quantities of flax are sown through all the county, and in some particular places a good deal is annually raised; especially in the parishes of East and West Monkland, at one end of the county, and that of Carnwath, and the parishes around it, at the other end. Much of the flax produced in the former, is very valuable. Flax growers rent land proper for it, at from 4 l. to 7 l. per acre; and there may be about 200 acres thus occupied annually. Women who purchase this flax, spin from three to five spindles of yarn out of a pound *, which they sell to manufacturers, to be wrought up into lawn and lace thread. The flax of the latter is of a coarser quality, and is spun by the women in the farm-houses and villages, into very useful yarn, from one spindle to half a spindle per pound, large quantities of which are sold in the markets of Lanark, Carnwath, Biggar, &c.

The spring seed time is very uncertain, depending on drought occurring, sufficient to dry up the winter's moisture. It is sometimes begun about the end of February, and sometimes scarcely finished by the first of May. Potatoes are planted from the middle of April to the middle of May, principally in drills made by the plough, from two feet six inches to two feet nine inches asunder. There have been instances of upwards of 24 tons of potatoes being taken from an acre; but the produce is frequently below half that quantity. It has been observed, that the more frequently potatoes are returned on the same ground, the produce is the less. The disease called the curl, also frequently occasions a great deficiency in the produce on the lower grounds, it not yet being prevalent on the heights, though it seems to be creeping upwards. No satisfactory discoveries having yet been made of the cause or cure of this disease, it must be left for future inquiry.

Barley is sown from the middle of April to the end of May. Lincolnshire and Siberian barley were introduced into this county for seed, and were somewhat in vogue for a time; but both these and the Tartarian oats, which were also pretty frequently sown, seem now to be entirely laid aside. The only kinds of seed now used are, the barley with

* A spindle is a parcel consisting of 48 cuts; each cut is 120 threads round a reel of two and a half yards circumference.

two

two rows of grains in the ear, and the Bear, Big or Chester barley, as it is called in different places, with four rows, both of which are well known. Barley is sometimes steeped to forward its growth; but when drought continues long after sowing, if the land is not well pulverized, the seed is in danger; its roots, springing with the moisture it contains, are withered and die. From 8 to 10 pecks are sown on an acre; and as many bolls are reckoned a good crop.

Turnips are sown from the end of May to the 10th of July; and, in dry early land, sometimes later. The culture of turnips was for a long time mostly confined to the light soils in the upper parts of the county; but of late it is so generally diffused, that there is scarce a farmer in the county who has not annually his turnip field, notwithstanding the disadvantage of a stiff cohesive soil and impervious bottom with which many have to contend.

Carrots too, are successfully cultivated on some favourable soils. The soil esteemed most favourable, is one that is open and friable with a dry sandy subsoil. The carrot seed is sown about the first of April, on ground which has been well dunged the former year for turnips or potatoes. The ground, after having been prepared by previous ploughings and harrowing, is laid in two feet ridges, and the seed deposited, by the hand, in a small groove made in the middle of each ridge. To prepare the seed, it is mixed with four times its bulk of fine sand, in a shallow box or vessel, three weeks before sowing. The seed thus mixed is kept always moist in a warm place, that it may begin to germinate before sowing. Fifteen ton of carrots are obtained from an acre, and have been sold in the neighbourhood of Glasgow, at 3*l.* per ton.

Rye has also been found to be very productive on dry sandy soils. It is commonly sown after potatoes, in the month

of

of October, as soon as the potatoes are removed. From half a boll, sown on an acre, 24 bolls has been reaped, on ground which would have yielded a very slender crop of any other grain.

Few field cabbages, or greens for feeding cattle, are yet cultivated.

The grasses cultivated are, red, white, and yellow clover, rye-grass, and rib-grass. The seeds of the *holcus lanatus*, and of some other native grasses, either saved in the field, or collected in hay-lofts, are sometimes sown instead of rye-grass. Sometimes only red clover is sown along with a little rye-grass; sometimes a mixture of more, or all of the above, are sown on the same field, according to the purpose for which it is intended. Grass-seeds are either sown among young wheat in the Spring, or along with oats or barley. Grass after wheat generally succeeds the first year, but is better the second year when sown with barley. In some parts of the county, it is observed, that clover is surer on a field from which two successive white crops have been taken, than where there has been only one.

An observer residing in a distant province, mistaking what is here advanced as a fact, to be laid down as a principle, says, this doctrine will not go down in these days. Though it had been observed, in many instances, that clover not only sprang more regularly, but grew more luxuriantly, after two successive white crops, than after a green and a white one, the assertion was made with caution, lest it might not always be the case: but the Author has had some strong instances of it under his eye this season, and is persuaded it is invariably so. If the observer, therefore, is dissatisfied that Nature does not act according to his system, he may change her operations if he can.

Of Plants not commonly cultivated.—The public could
derive

derive no instruction from any history which could here
be given, of the culture of plants not commonly cultivated, there not being accounts sufficiently authentic, to
determine whether such plants have failed of being more
extensively propagated, from their not being adapted to
the soil and circumstances, or from want of due attention to the culture.   There is no class of men less fit for
new and curious inquiries than husbandmen.   Habit has
confined them within a routine which admits of few material variations; and both mind and body are so much
occupied with their ordinary labours, that neither leisure nor disposition are left to look after new ones.   It is
only among people who follow agriculture, either for
amusement or partial employment, that the existence of
that inquisitive turn, which leads to discovery, is to be
found.   But these people are frequently too sanguine,
not always happy in choosing the subjects of their experiments, and often diverted, by other circumstances,
from pursuing them to the final point.   Hence the culture of different plants has been introduced, and dropped, without it being possible to decide, all circumstances considered, whether the more extensive culture of
such plants would have been beneficial or not.   It may
be said, however, that since the introduction of artificial
grasses, potatoes, and turnips, the culture of no plant of
general utility has taken place in this county; and from
thence an inference may be drawn, that every kind of
culture, of such description, will, at length, find its way
of itself.

ART. 2. *Rotations of Crops.*—Rotations are as various
as the climate and the soil.   It will be sufficient to mention a few, in different parts of the county.   Among the
light soils of the upper ward, the two following are the
most prevalent: by the first, the whole arable land is

I 3   divided

divided into eight parts, and each in its turn undergoes
the following rotation : 1*st* year, fallow, or turnip in
drills and dunged, and a portion in potatoes ; 2*d*, barley,
and sown with grass-seeds; 3*d*, hay ; 4*th*, 5*th*, and 6*th*,
pasture ; 7*th* and 8*th*, oats. According to the second,
the half, or as much of the farm as is judged convenient, is laid out in four divisions, each in its turn managed as follows : 1*st* year, fallow or turnip, &c. 2*d*, barley or oats, with grass-seeds ; 3*d*, hay ; 4*th*, oats. The
remainder of the farm lies in grass, and is pastured by
the dairy cows, cattle to be fattened in Winter on turnips, &c. A part of this is taken in at pleasure, in exchange for a part of what has been kept in culture. In
the light lands, in the lower part of the county, turnips are
not cultivated; and there different practices prevail. The
following rotation is the most approved : the farm is
divided into five lots, each managed thus ; 1*st* year, the
land is spring-fallowed, well manured, mostly with
Glasgow dung, and potatoes planted in drills, and kept
clean by summer hoeing ; 2*d* year, wheat sown as soon
as the potatoes can be taken up, and grass-seeds sown
among the wheat, in the Spring; 3*d* year, hay, twice
cut; 4*th* year, hay, once cut, and the after foggage pastured; 5*th* year, the field having been manured with a
compost of lime and some kind of earth, in the go harvest, is cropped with oats. Sometimes barley is sown,
instead of wheat, the second year ; and sometimes the
land is spring-fallowed after the wheat and barley without manure, sown with grass-seeds ; and thus the rotation takes in another year. In the clay soils, in the same
neighbourhood, the land undergoes a clean fallow, and
is strongly dunged for wheat. Beans succeed the wheat;
oats, with grass-seeds, the beans, &c. Among the variety of practices which obtain in the middle country, the
two

two following are prevalent : by the first, the land is summer-fallowed, dunged and limed, and wheat sown; beans and pease succeed the wheat ; and oats, with grass-seeds, the pease and beans ; the grass is cut for hay, one or two years, and afterwards pastured so long as it is thought proper to let it rest. By the second, the land is dunged or limed, or laid over with compost upon the grass, and cropped, 1*st*, with oats or pease ; 2*d*, with pease or oats ; 3*d*, with oats, sown with grass-seeds. It is the rule, in some cases, to let the land rest as many years as it has been cropped ; in others, double that time. On the more elevated land, on the east and west boundaries of the county, circumstances are unfavourable to the rotation of green and white crops. There some kind of manure is laid upon the turf, two or three crops of oats taken, and grass-seeds sown with the last. Some landholders have bound their tenants, never to crop more than a fourth part of the farm, the rest being occupied in pasture or hay. Some tenants, who are not bound, follow nearly the same plan, from conviction that it is best in their situation.

SECT. V.——OF GRASS GROUNDS OF ALL KINDS, AND THE PURPOSES TO WHICH THEY ARE APPLIED.

Grass grounds are naturally divided into two kinds, viz. 1*st*, Such as from situation and circumstances are deemed unfit for any other purpose; 2*dly*, Those which are laid in grass, not only for present profit, but also in order to render them more profitable in tillage, when they shall be used in that way. The first kind must chiefly be used in rearing live stock ; and, there-

fore,

fore, what relates to rearing may be included under that
article. The second is more frequently used to supply
food to stock already reared, and fatten animals for the
table; what relates to the provisions derived from animals, will, therefore, come under the second article.

ART. 1.—*Pastures that are deemed unfit for any other
purpose.*—Among the mountains, in the upper part of
the county, flocks of sheep are kept. About 50 years
ago, corn was cultivated on some of the lower grounds,
at the feet of the mountains, more than sufficient to
feed the inhabitants; and many neat cattle were kept.
At that time, the sheep, more restrained on the summits
and poorer pastures, on many of which an animal can
hardly live in a storm, were small, and of no great
esteem in the market: but a succession of bad seasons
ruining the crops, has obliged the farmers to abandon
agriculture; and torrents, from time to time, having
brought down stones from the hill sides, and choaked
up the channels of the rivers and brooks, many of the
little valleys are so frequently overflowed, that they are
no longer capable of cultivation. Scarce any corn is
now raised, few black cattle are kept, and the sheep,
now indulged in the best pasture, have increased in size.
They bear very coarse wool; but are extremely hardy,
and much esteemed by the Yorkshire dealers. They are
now reckoned fully equal to the best in the neighbouring counties. The flocks of the farmers formerly consisted of about 1000 sheep each; but of late some individuals rent more land, and are possessed of from 3000 to
7000.

The despair produced by such natural obstacles against
every attempt towards the improvement of the country,
has, perhaps, been too easily admitted. Two different
causes have contributed to strengthen and confirm this
despair;

despair : the effect which the arguments of philosophical economists has produced on the conduct of landholders, and the invitations which rising manufactures have held out to the people. But one instance of land cultivated here, probably of greater altitude than any other in the island*, will show, that if the bent of the stock, and industry of the society, were more applied to the improvement of their most important and permanent property, much might be done every where to increase fertility, even in the most unpromising situations.

The inhabitants of Leadhills find it necessary for their accommodation, to keep milch cows, which go to pasture in the neighbouring moor. The dung of these cattle is laid upon barren moorish ground near the village ; this ground is levelled, formed into narrow ridges with the spade, and potatoes, turnips, cabbages, greens, &c. cultivated upon it, by the inhabitants at their spare hours. One spot has been laid down in grass, and another taken up, from time to time ; and there are now a good many acres, naturally of the most barren soil, bearing grass little inferior to that on the best land in the county. Besides the vegetables above-mentioned, and green food in summer for a great number of cows, 9000 or 10,000 stones of hay are annually made, for winter feeding.

Along both sides of the county, so far as the moors extend, sheep pasturing is followed ; and the quality of the sheep is in proportion to the pasture. Where the moorish pastures are disjoined by the intervention of arable fields, the flocks are smaller, from 300 to 400, and a greater number of black cattle are reared. On the

---

* The height of the mountain Tintock has been stated at 2260 feet above the level of the sea, and the site of Leadhills at 2000 feet ; but some geometricians assert, that they have found the height of Tintock to be 2400 feet, and that Leadhills is nearly on the same level.

coarse

coarse high land, on the east side of the county, it has
been found the most profitable practice to winter young
black cattle. The pastures are allowed to grow untouched, from the end of May to the end of August.
The rushes, and such other coarse herbage as grow on
the marshy places, are mown, while tender, and laid up
for winter food. About the end of August, the pasture
is stocked with small young Highland cattle. They live
upon the grass, when the weather is moderate; when a
storm happens, the winter fodder is given them on some
sheltered spot. When they can be accommodated with
some slight shed, it is reckoned an advantage. The pasture necessary for wintering one of those, is thought
equal to that of five ordinary moorland sheep of the first
year (*hogs*). The cattle are sold off in May, and are
generally increased in value from 25*s*. to 30*s*. per head.
In most of the farms in the elevated parts of the country,
where a less proportion of the land is tilled annually, and
the pasture less rich than in the lower parts, it is the
practice to breed a good many neat cattle, nearly about
the half of what are brought forth; mostly females, few
oxen being reared in the county. These feed on the
pastures where they are bred, and either supply the place
of old cows disposed of, or are sold at three years old,
with the first calf, at from 4*l.* to 7*l.*, and since this Report was first printed, from 7*l.* to 10*l.* a head, according to
their size and figure. In the lower parts of the country,
fewer calves are reared; sometimes no more than necessary to keep up the stock of milch cows. Most of the
farmers through all the county rear young horses, chiefly
of the draught kind. In some cases, one foal, in some,
two are reared, on a plough-gate annually, according to
the quantity of pasture that can be spared, for feeding
nursing mares and colts. Those who summer-fallow
                                        much

much of their farms, and have much manure to carry from a distance, rear no foals, and prefer geldings to mares, for performing their labour.

The natural meadows, repeatedly mentioned, are saved always for hay. The bottom being cold and swampy, the grasses spring late in the season, and the hay is not cut before the beginning of August. The situation being damp, and the hay soft, it is frequently got with difficulty, and suffers in the making, both in colour and flavour. About 50 years ago, this was almost the only hay in the county, and the stables of inns, &c. were supplied with it; but since the culture of artificial grasses became general, it is no longer a subject of commerce, and is consumed by the live stock on the farm.

ART. 2. *Artificial Grass, Dairy, and Feeding.*—The first crop of sown grass is generally mown for hay. The hay harvest commences about the beginning of July. Haymaking is conducted different ways. Some new haymakers mow only when the weather is fair and the grass dry. Rakers immediately follow the mowers, and put up the hay into very small cocks; it is afterwards turned daily, the size of the cocks being increased as it dries, till it be ready for treading up in field ricks. However this practice may succeed in the southern counties of England, it does not seem to be well adapted to the state of this country. Unless the weather be uncommonly favourable, the hay always suffers somewhat in colour; and under repeated rains is greatly injured. Many of the most expert husbandmen now follow a very different mode. The hay is allowed to lie in the swathe for three or four days, and then, if the crop is not heavier than from a ton and a half to two tons per acre, it is raked together, in a dry time, and immediately tread up into round or oblong ricks in the field, where it stands till it

is

is thought to be in condition to be put safely into large stacks. Very heavy crops must be turned in the swathe, and both sides exposed to the drought; and when great rains have occurred, all the hay must be turned. What lies immediately under the shade of trees or hedges, is removed towards the middle of the field, and allowed to lie open to the drought for a few hours. In this cheap and simple manner, hay is better preserved than by more operose methods.

Few have attempted to save clover seed; but the seed of rye-grass is frequently saved. When this is to be done, the hay is bound in sheaves after the mowers, and set up in shocks for a week; the seed is then thrashed, or skutched off, and the hay put in the ricks.

For the most part, the latter growth, after the hay, is pastured either by the live stock on the farm, or cattle bought for fattening; and as there are but few instances where a second crop of hay is taken, the grounds which have been laid in grass are pastured the same way, so long as they are allowed to rest, after the first year.

It was a prevalent custom in this county, to keep a great many milch cows, long before the profits of the dairy became considerable, or the management of it was understood, it being thought necessary to keep constantly a number of cattle for making dung to recruit the arable land. These were led around the indifferent pastures, in summer, travelling a great deal to gather a moderate support. In winter they were driven out to the fields, to endure all the vicissitudes of the weather, through the day, and fed at the stall, in the evenings and mornings, with a little straw; and were generally much reduced before the return of the grass. With this treatment, the quantity of milk which they yielded was little, and

as

as neither the demand nor the price was considerable, the management of the dairy was little studied.

The feeding of calves was the first object of profit. On the elevated lands, where the harvest is less perfect, part of the unripened oats were taken to feed the milch cows, which increased the quantity and richness of their milk. The calves which were brought forth about the latter end of harvest or beginning of winter, were fed at first with the milk of their dams, and afterwards with the thicker milk of those which were beginning to dry, having been taught, from the first, to drink all that they got. In this manner, rich veal was fattened and sent to Edinburgh, from Christmas, onward, to supply the tables of the wealthy, where Lanarkshire veal has been long famed, and bought at exorbitant prices. In the progress of improvements in the country, a better provision of winter food for cattle has been made, and, by feeding milch cows with potatoes, turnips, &c. the practice of fattening veal has been much extended of late; so that, though the increase of wealth and luxury has greatly enlarged the demand, the rise of price, on this fine veal, has not been in proportion to that on other kinds of provisions: however, in all places distant from markets, it is still considered as the most advantageous way to dispose of winter milk.

As the prices of butter and cheese advanced, the owners of dairies, by degrees, became more studious, not only to increase the quantity, by paying more attention to the feeding of milch cows, but to ensure a preference in the market, by adapting the quality to the taste of the best customers. This was, however, confined for a time to the most favourable situations; those who were less happily situated, attributing solely to soil and circumstances, what was in a great measure owing to skill, attention,

tention, and cleanliness, were not eager in imitating, when they despaired of arriving at perfection. It was not till after the first peace of Paris, when the rapid progress of commerce and manufactures brought a new influx of wealth among the inhabitants, and greatly increased the demand and price of butter and cheese, that attention to the dairy became general. Fresh butter especially became an indispensable article at every breakfast table, and all that was made with care and cleanliness, sold quickly at high prices. This stimulated the country house-wives, throughout the county, to study and follow the requisites of the dairy; and now all the different articles which it produces are to be found, of an excellent quality, even in the elevated parts of the county, which were formerly deemed the most unfavourable; particularly within the reach of Glasgow, where the opportunities of comparing, in a great market, prompts people to observe all the minutiæ necessary to bring their commodities to the greatest possible perfection. A pound of butter, indeed, made on the high pastures, is supposed to yield a little less oil than one made on the low; but the taste and flavour is equally good, and the former is less apt to become rancid in keeping. The produce too is less in proportion, on the former than on the latter. The dairy business is conducted three ways in this county:—either the whole milk is made into cheese, or butter is made of the cream, and cheese of the skimmed milk; or, in the most populous parts of the county, where there is a great demand for butter-milk, as an article of food to the poor, the whole milk is churned. People are induced to adopt any one of these ways, either from situation, or from some circumstances of supposed conveniency. There are some pastures from which the milk yields proportionally more

butter,

butter, others more cheese; but there are very few instances in which this variation in the nature of the pasture
has been found considerable, and there is no sure rule
of judging but experience. It is mostly fancy which de-
termines the choice, the profit from the different ways
being nearly equal in general. By averages made up
from the reports of the owners of dairies in different
parts of the county, eight Scotch pints of new milk, or
the cream taken off it, will produce a pound (or 22
ounces avoirdupois) of butter; 80 pints will therefore
produce 10 pounds, and after the cream is taken off for
the butter, 70 pints will remain for cheese, and this will
make a stone (22 pounds avoirdupois) of cheese, saleably
dry: about 53 pints of new milk will make a stone of
cheese. The state of the different practices then will
stand thus:

80 Pints churned.			*New-milk Cheese from 80 Pints.*			*Butter and Skimmed Cheese from 80 Pints.*		
	s.	*d.*		*s.*	*d.*		*s.*	*d.*
10lb. of butter, at 8d.	6	8	24lb. at 5d. -	10	0	10lb. of butter, at 8d.	6	8
76 pints of milk, at ½d. per pint,	4	9	60 pints whey, worth 2d. per gallon,	1	3	1 stone cheese,	4	8
						6 pints butter-milk, at ¼d. per pint,	0	4½
						50 pints weak whey, at 1d. per gallon,	0	6¼
	11	5		11	3		12	2¾

Two or three pigs are frequently reared on the whey at
a dairy, and fattened with potatoes; but the feeding of
hogs is not carried to any great extent in this county.

Of two cows, of the same size, and perhaps from the
same parents, one often gives a good deal more milk
than another, and the milk of one cow is of a richer
quality than another: but cows in general give more or
less milk, and better or worse, in proportion to the
quantity

quantity and quality of their food, the regularity with
which it is administered to them, and to the ease they
enjoy. It has been always observed, that two or three
cows, kept by themselves, are more productive, in proportion, than greater herds, though the few should enjoy
no visible advantage over the many, but perhaps the
contrary: which seems to show, that undisturbed quiet
and minute attention, are of great importance. Particular instances might be given, of cows in this county giving
a great deal of milk; but it is not from uncommon instances, but general averages, that a just idea of a district
can be formed. It is reckoned a moderately good milch
cow, that gives eight Scotch pints a day; and supposing
her whole milk through the season to be equal to 150
days at that rate, it will make 150 gallons, which, valuing
every ten gallons at 11s. 3d. the lowest of the above states,
it will make the annual produce of a milch cow, exclusive of the calf, 8l. 8s. 9d. This, however, is far above
the average produce of the dairies in this county, which
runs from 6l. to 4l. per cow, according to the situation;
and about one-third more may be added to this and the
other money articles, on account of the rise of price in
commodities of the dairy since this Report was first
printed.

Since this Report was first published, the prices of the
produce of the dairy are greatly advanced; butter has sold
in Glasgow market at 1s. 7d. per pound, and may be stated
at an average through the county at 1s. 5d.; new milk
cheese at 10s. 6d. per stone; skimmed milk cheese at 8s.;
butter-milk from the middle of September to the first of
May, sells at 1d. per pint. The average produce from
each milch cow may now be from 8l. to 10l. yearly; and
in the neighbourhood of towns, still more, the breed of
cattle and the manner of feeding having been gradually
                                        improving.

improving. The money value of milch cows has been falling since spring 1804, and is now 25 *per cent.* lower than it had been for some time before that period. The value of articles of the dairy also suffered a sudden depression in the beginning of the month of April 1805, the price of butter having fallen at Glasgow, in one market day, from 1*s.* 7*d.* to 1*s.* 1*d.*, but is now again upon the advance.

Those who keep milch cows give them the best pasture in summer. When the grass fails, wherever turnips are raised, these are given, in order to protract the milk season. But full feeding with turnips renders the milk nauseous; and none of the recipes which have been recommended can cure its rankness. When cows get plenty of good fodder, and are only partially fed with turnips, the purpose of continuing milk is, in a great measure, answered, and the effects on the flavour of the milk little perceptible. If turnips be washed, or if the weather be rainy when they are taken up, the flavour of the milk is the stronger. In some places potatoes are used in place of turnips, for the same purpose, and are much preferable, producing not only plenty of milk, but of a rich quality and of a good flavour*.

---

* Though the profits of the dairy are moderate, it seems to be very doubtful, if a farmer can make as much of his pasture by any other management. It was a late subject of debate, proposed for the consideration of a respectable society of practical farmers in this county, whether keeping milch cows, fattening cattle, or fattening sheep, was the most advantageous; and it was decided in favour of milch cows, by a great majority of votes. The principal arguments on this side were, that the produce of the dairy was never equal to the demand, and therefore the market less fluctuating; that the farmer who kept milch cows had little occasion to go around the country to markets, and had more leisure to attend to his farm; that by keeping milch cows, properly fed, attended, and littered, the greatest quantity of manure, for the improvement of the land, could be made.

CLYDESDALE.]　　　　　　K　　　　　　Sheep

Sheep and neat cattle are fattened for the table, both in summer and winter. A good deal of enclosed land, not in lease, is occupied for this purpose, and is either stocked by the proprietors, or by jobbers who rent it by the year. Some farmers employ a part of their enclosed land the same way. The ground is stocked, either with wedders, or ewes with lamb, or with neat cattle, according to the suitableness of the ground and the fences, and to the opinion of the grazier. Sometimes the beasts are bought in, about go harvest, and wintered upon the ground; sometimes the pastures are kept void in winter, and stocked after the grass has got up. In winter, both sheep and neat cattle are fattened upon turnips in the upper ward, where the turnip husbandry prevails most. Sheep are either enclosed in nets, and fed upon the turnip ground, or, if it is not dry enough, the turnips are carried to an old pasture field and given them. In this last case, only three-fourths of the turnips are taken up, and the sheep are afterwards allowed to stroll over the field and gather the remainder, that some manure may be left. Neat cattle are fed at the stall. Any turnips that are raised in the lower parts of the county, are mostly used in feeding neat cattle. Potatoes have been used with great success in feeding cattle. They are given either raw or boiled. Some think the latter has the quickest effect. The same root has been found a very frugal and a very hearty food for working horses.

CHAP. VI.

GARDENS AND ORCHARDS.

———————

LEAVING ornamental gardening, and those niceties of the art, practised to produce the delicate fruits which the want of sun denies us naturally, it will be sufficient, for a work of this nature, to mention the gardens culti-vated for supplying the more simple and urgent wants of man. The chief of these are the mail gardens around the city of Glasgow, from which that populous place is sup-plied with all the variety of culinary vegetables produced in this country, at their different seasons; and though the first articles do not come so early to market as at Edinburgh, where the soil is light and dry, as good sale always stimulates the exertions to answer it, the growing wealth of Glasgow provokes the gardeners to make quick advances. So plentiful is the supply, that much garden stuff, towards the latter end of the year, is sold at a cheap rate, and carried to the neighbouring villages. Families in the country, and many families in the smaller towns, are well supplied from their own gardens. And the gardens round the villages in the country, before mentioned, afford great accommodation to the inhabitants, as well as whole-some and innocent recreation to those of sedentary em-ployments.

The Clydesdale orchards lie mostly between the bottom of the lowest fall of the river and the mouth of the south

Calder.

Calder. They are chiefly of apple trees, with a mixture of pear ones, and some of plumbs. Cherries are more rarely cultivated, being much subject to the depredations of birds. Few of the orchards are large, but many small ones are planted up and down the country. They were stated, in the former Report, to amount to 200 acres; but some new ones having been lately planted, and some more minute information having been since received, they may be safely said to be upwards of 250 acres. And since this Report was last published, the planting of orchards has gone on so rapidly, that the orchards of the county must now exceed 340 acres. The produce is very precarious, the fruit being frequently destroyed in the blossom, by spring frosts and caterpillars. Some years, the whole value has amounted to upwards of 2000*l*. And in the years 1801 and 1804, by the best computation that can be made, the value of the fruit from the different orchards exceeded 5000*l*. each of these years. But this was not so much owing to an increase of fruit from the orchards lately planted, few of them having arrived at any perfection of fruit-bearing, as to a gradual rise of the price of fruit, and both those years being productive ones. A remarkable instance is mentioned of the fruit produced on half an acre of ground, in the former year, bringing 150*l*. to the dealer who carried it to market. The value of the fruit is not always in proportion to the number and size of the trees. Those who cultivate the ground around the trees, taking care not to injure the roots, and giving manure from time to time, have finer fruit, and a much greater quantity, in proportion, than those who do not. Much also depends on adapting the trees to the soil and exposure. Though the different kinds of apples, &c. are generally engrafted on the same kinds of stocks, each assumes the habits peculiar to the scion.

scion. Those who have been attentive in observing this, and choosing the kinds best adapted to their situation, have found their account in it.

But it ought not to be understood that the choice of the stock is of no importance. Native crabs are the hardiest, and prove the most durable trees. Codling stocks, and those raised from the seeds of good fruit, generally produce also finer fruit; but the trees seem to be more subject to disease. Yet diseases are perhaps as often communicated from the scion as from the stock. But the causes which produce the phenomena occurring in the orchard, are so intricate and incomprehensible, that the most attentive and acute cultivator can neither avert the injuries and maladies to which the trees are liable, nor cure those that are diseased. There is one disease to which fruit trees are subject, here called the canker. It begins with a blotch in the bark, which gradually spreads, and produces a mortification in all the infected members, and finally kills the tree. In some soils and situations apple trees are most infected with it; in some, pear ones; in others it indiscriminately attacks both. Again, there are soils on which the trees grow with vigour for some time, till a decay, beginning at the extremity of the branches, gradually proceeds towards the trunk. When the diseased branches are lopt off, the tree sometimes recovers, and puts forth new shoots. The pith of those branches becomes black and corrupt before vegetable life ceases in the exterior parts. When any part of the infected pith is left in lopping, the disease proceeds. But frequently all lopping is ineffectual, and the tree dies, sometimes suddenly, and sometimes pines away by slow degrees. Many varieties of the different fruits are cultivated, distinguished by particular names, which are only local and provincial. Some of

K 3 those

those varieties, which prove healthy trees in one situation, are infected with the above diseases in another, in
which other varieties enjoy health. Hence the congeniality of the soil, and exposure to the peculiar nature of
the tree, and the contrary, is also a cause of health or
disease. But there is no general principle to direct the
cultivator of the orchard. All must depend on a long
course of topical experience, by which the kinds of fruit
trees, which have been found to thrive and bear best in
any particular spot, may be preferred.

The Clydesdale orchards are mostly planted on steep
hanging banks; and on such they have been found to
succeed better than on plains, as subterraneous water flows
most quickly away. Most of them stand on soils pretty
cohesive, and on such the trees have been supposed to
be surer bearers than on open sandy soils: yet there are
instances of orchards, on friable and gravelly soils, uncommonly productive. The apple tree in general succeeds
on a pretty hard soil, provided the bottom be dry; but
when the roots penetrate a subsoil holding stagnant water,
or greatly charged with the orange oxid of iron, commonly called ochre, the tree fails. The pear tree requires a soil of greater depth, and more soft and moist;
and will thrive in an ochery subsoil where the apple tree
fails. It lives to a greater age, and arrives at a greater size,
and more towering height, than the apple tree. The plumb
tree does not succeed in the very stiff cohesive soils. It
requires a considerable depth of dry friable mould. All
the fruit trees which have been engrafted are more delicate than those in a natural state, and require a more
attentive culture. But the cultivators here differ in opinion, respecting the degree of attention. Some are for
digging over the ground every year, at least around the
roots of the trees as far as the branches extend, dunging

ing at intervals, and keeping down all weeds and herbage during the summer with the hoe; others think the trees are injured in their health by the soil being kept too loose and open, by perpetual stirring and dunging, and prefer digging and dunging at intervals, and then letting the ground rest in grass for a few years. Some again prune the trees, taking many branches out of the centre, to give free admission to the light and air; some approve not of pruning, except cutting out those branches which bear upon, and offend others; and others would not deprive a fruit tree of a single branch. Plumb trees are generally planted round the verge of the orchard, and are profitable, not only for the fruit they bear, but for sheltering the other trees. All fruit trees require shelter, and do best when they are embosomed in woods. The latest kinds do not arrive at perfection in backward seasons, and therefore it is always proper to have a good mixture of summer and autumnal fruit along with the winter's. A just proportion of apples, pears, and plumbs, is also proper, as one kind sometimes succeeds when another fails. Upon the whole, though the produce of the orchard is precarious, when the original insignificance of the grounds on which fruit trees succeed, is considered, and the ready sale and high price which the manufacturing towns afford for fruit, an orchard planted with judgment, and carefully cultivated, is certainly a profitable kind of agriculture. The depredations committed on the orchards are become more frequent and daring as the manufactures of the county have increased, and are a great discouragement to this species of cultivation, particularly that of small orchards, which cannot defray the expense of watching in the night.

Besides the larger fruit, great quantities of gooseberries and currants are cultivated, and, when well managed,

K 4

naged, are said to pay very well. The gooseberry and currant trees are dug around annually, kept on a single stem, and dunged every second year. Many new varieties have been lately introduced, of a large size but inferior flavour to the old ones. Vast quantities are every year brought to market, to the value, it is supposed, of one-third of the large fruit.

CHAP. VII.

WOODS AND PLANTATIONS.

THIS Chapter may be divided into two sections; 1st, The kinds of trees, and extent of plantations; 2d, Observations on the different kinds of forest trees.

SECT. I.—KINDS OF TREES, EXTENT OF WOODS AND PLANTATIONS, &c.

THERE are scarcely any instances of spontaneous coppices above the uppermost fall of the river. But some of the principal landholders, of late, have done much to adorn the country with planting. In the early part of the last century, except a few trees about some of the houses,

houses, this part of the country was quite naked. There are now about 1800 acres planted, three-fourths of which, at least, has been done in the last twenty years. The trees are of various kinds, but the Scots pine and the larix are the most prevalent. From the top of the falls downward, coppices arise every where, near the sides of the river and the streams which fall into it. These consist of oak, ash, birch, elm, alder, holly, gean, or wild cherry tree, sallow of different kinds, &c. intermixed with hazel or other shrubs. Of these there are 760 acres in the lower part of the upper ward, besides 580 acres of planted wood, making the whole in this tract 3140 acres. In the middle ward there are 1350 acres of coppice, and 2830 acres of planted wood. There are few coppices in the under ward, perhaps not 40 acres altogether; nor is the planted wood of great extent. Hedge rows and narrow stripes surround the small enclosures, and give the country a clothed appearance, but probably the square contents do not exceed 700 acres. This makes the whole of the woods in the county 7990; but there is now reason to believe there are considerably above 8000 acres. For these three years preceding, as well as the present year 1805, large additions have been making to the plantations of the county; but they have been so generally diffused, that it would be impossible to make any tolerable computation of their extent.

The copse woods are sometimes cut once in 25 or 26 years, but are more frequently allowed to grow 30 years; and an acre is sold at from 20*l.* to 30*l.*, and now from 60*l.* to 70*l.* Woods that are extensive are divided into separate lots, called *hags*, one of which is appointed to be cut annually. These hags are from three to seven acres, according to the extent of the wood, and the sale in the neighbourhood. It has long been the custom to

leave

leave 20 or 25 select trees, called *reserves* or *witters*, in
an acre, at each cutting. The intention of this seems to
have been, to furnish purchasers with an assortment of
wood of different sizes. This practice is still continued,
but appears to be an injudicious one. When those trees,
drawn up long and slender, by the shelter of the surrounding wood, stand in an exposed situation, they are
unable to bear the blast after they are left single ; and if
they are not quite stunted, make little acquisition of size.
Should they happen to thrive, they do more injury to
the young growth around them, than all the additional
value they attain. An observer on this, who seems to
be a gentleman of good sense and candour, thinks, " that
" the reserves are so far distant from each other, as to
" cause slight injury to the surrounding wood, since the
" lower branches of the former might be lopped, to give
" the sun and air easy access to the latter ; and that it
" would be advantageous to have a supply of the various
" wood and timber which may be required upon the
" same spot." This is, therefore, here stated, and the
whole left to the judgment of the proprietors of woods.
The custom of leaving reserves, however, is now much
given up.

Formerly there was no kind of trees planted to any considerable extent but the Scots pine ; and there are still
more of this kind than any other, it being planted to
protect the deciduous kinds. When this is the case, it
ought to be cut down before the others grow too tall and
weak. When it is planted unmixed, it is reckoned the
best practice to put the plants pretty close together,
about 6000 to an acre; so that by the support they derive
from one another, they may grow up straight and tall,
and the tops meeting, may exclude the air, and smother
the under branches, while they are still small and weak.
                                                    This

This is called pruning themselves, and is found to be the best way for preserving the health of the trees, and obtaining valuable timber. It is absolutely necessary that open drains be made through all the hollows, that no water may stand. When the general height of the trees is about eight feet, and no living branches on them but at the top, it is time to begin to weed them; and this ought to be regularly carried on, according to contingent circumstances, for five or six years, till the trees stand nearly at such a distance as may give them room to grow to useful timber; all future weedings being dangerous, not only for opening avenues which may admit too strong a stream of air, but on account of the noxious quality which the putrid roots of fir trees, cut down after they have arrived to a considerable size, show, by frequently killing those which have been left standing. The abundance of coal and peat in this country, renders the first profits of planting inconsiderable. In the thinly inhabited parts of the upper ward, where there is little demand for small wood to make fences, &c. the first weeding of plantations is a heavy expense. Even wood farther advanced had little sale, till the erection of the iron works around occasioned a great demand for small trees, for supports in the mines, and for wood of every kind for different purposes. In the lower and more populous country, every kind of wood finds some market; and valuable timber of late has sold very high. The Scots pine planted on very poor land, 25 years old, has sold for 25l. now 40l. per acre; the same when properly thinned, and standing 50 or 60 years, for 80l. now 160l. and upwards. The prices of different kinds of well grown timber, per cubic foot, were as follow:

In

	In 1797.	1805.		In 1797.	1805.
	s. d.	s. d.		s. d.	s. d.
Pine or fir,	0 9	1 6	Lime (Linden),	1 4	2 0
Oak,	2 0	3 6	Poplar (white),	1 4	2 0
Ash,	2 0	3 0	Birch,	1 0	1 0
Elm,	2 0	3 0	Holly, for ve-neering,	5 0	5 0
Plane,	2 0	3 6			
Beech,	1 6	2 0	Gean-tree,	1 6	2 0
Sallow, for mill-timber,	2 6	2 6	Apple and pear-tree,	1 6	1 6
			Horse-chesnut, for veneering,	....	3 6

SECT. II.—OBSERVATIONS ON THE DIFFERENT KINDS OF FOREST TREES.

IMPORTANT instructions, with respect to the kinds of trees adapted to the soil, bottom, and exposure, intended to be planted, may be obtained, by observing the different degrees of success attending those trees in the numerous plantations of this varied county. It may not be improper, therefore, to insert what occurs in different situations, respecting the different species usually planted.

The Scots Pine, or Fir, as it is usually called, is not adapted to the greatest altitudes. It begins to shoot in April, and completes its year's growth by the middle of June; so that the winter often continues to reign in the heights, where the sun is not reflected till the summer of the pine be over. Unable to struggle with a repetition of such seasons, it languishes and dies. Of all the trees of this kind, planted at Leadhills something more than 40 years ago, and very carefully treated, only two or three remain, in a very sickly state. On a moorish hill in the same neighbourhood there is a plantation standing

standing a little higher, and the soil incomparably worse. The trees are now about three or four feet high, and have already ceased to make farther progress. Besides the great elevation, there is perhaps something in the bottom injurious to this plantation. The Scots pine planted in much lower situations upon a dry shivery whin rock, the parts of which are separated by ochery fissures, ceases to vegetate in a few years, though the thin soil on the surface be good. There is also a kind of laminated clay, much disposed to dissolve with water, not favourable for this, nor any of the pine tribe. It succeeds very well, however, in most parts of the clay ground in the country, if care be taken to prevent stagnant water. It does exceedingly well on land covering the free-stone rock; but the best timber is produced on hard dry gravelly soils.

The Siberian Pine, the Cedar of Lebanon, and some others of a similar nature, have been introduced with very little success. The short intervals of mild weather which happen in the beginning of the spring, excite them to vegetate too early, and the next cold blast destroys the young buds. The New England pine thrives, in a tolerable soil, from 12 to 20 years, in proportion to the exposure, after which it generally begins to decay. The longest standing, and the largest, are on the low grounds near the house of Dalziel.

The Spruce is also unfit to weather the storm, on the greatest heights. It succeeds on the hard dry rock, where the Scots pine dies, but frequently decays at the end of 18 or 20 years, on stiff wet clay. Its most favourite soil is that which is dry and gravelly.

The Silver Fir thrives in clay soils, where the spruce fails; nor is it averse either to the hard rock or gravelly soil; but it makes no progress on any soil that is very poor.

poor. Unfortunately it frequently suffers severely from the frosty mildews of the spring, particularly in its youth.

The Larix is now found to be the most hardy alpine plant. In most places it makes greater progress than almost any other tree; and there is scarcely any soil, that is not drowned with water, on which it will not succeed. It suffers most in too luxurious situations, where its soft shoots, unable to keep erect, bend away from the slightest gale.

The severe frosts in May 1802 and 1803 succeeding some mild weather in the month of April, which had brought on the vegetation of the larix, killed many of the young plants of this kind, particularly those planted in low sheltered situations, and deterred people from continuing to plant this tree; but such as had escaped having continued to succeed, it is again become a favourite plant. It sometimes is seized with disease, and dies when placed on miry clay.

The Birch is next to the larix in the progress of its growth, and equal to it, in ability to stand the blast in alpine situations. The birch is superior in the plain. But in whatever situation it is placed, it delights most in a light soil and dry bottom. Notwithstanding, it thrives in moist soils, with very moderate draining.

The Ash, when it enjoys a sufficient depth of good soil, is capable of braving the storm, and pushing up its head, in the most exposed situations; but in a thin soil covering a stiff argillaceous bottom, it can make no progress. It thrives well, however, in marshy soils where the banks are steep, so that the water flows away without stagnating. This is perhaps the most important wood in the country, being useful in all its ages, and for most purposes.

The Mountain Ash is a hardy native, which grows
freely

freely in almost all soils and exposures; but its favourite
situation seems to be in hanging banks among woods and
coppices. It and the gean-tree, or wild cherry, propagate themselves much when left at liberty, but putting
up suckers from their roots.

The Beech comes near to the ash in capacity of
braving the storm, and has much the advantage of it in
thriving in poor or stiffer soils; but there are some barren argillaceous bottoms too much even for the beech,
and it is most successful in friable soils. Its young shoots
are soon affected by frost, but it quickly recovers.

The Sicamore and Elm require a light soil and a dry
open under-stratum; and when this is the case they
thrive in a situation pretty much exposed.

The Oak is less patient of the blast than most of the
trees of the forest. Being late in putting forth its leaves,
it continues to grow till the season is far advanced; and
the immature wood of its late shoots, unable to resist
the piercing effects of the cold wind in exposed situations, withers before the next spring; so that, like Pe-
nelope's web, the progress of one season is undone in the
following. The most favourable situations for the oak,
therefore, are hollows or hanging slopes, where the
sharp winds are broken by the neighbouring heights.
In such situations, if stagnant water be avoided, it will
thrive in the stiffest soils, and with its strong roots penetrate the densest bottoms. Some plants of this kind form
the bud on the top of the branch, preparatory to the following year's growth, early in autumn, and are less subject
to be injured by the winter's cold. It is doubtful whether this be a seminal distinction or an accidental variety;
but such forward plants have always the bark of their
young shoots of a dull purple colour. In planting oaks,
it would be proper to choose these till the matter be better

ter known. The oak tree in coppices, and also the mountain ash, is valuable, on account of the bark; much demanded for tanning leather, and now is sold at 14*l.* per ton. The bark of the sallow and birch are used for the same purpose, and sold at half that price. The value of oak timber is well known. Though its growth is slow in infancy, when it is placed in a favourable situation, it will make a progress in the course of 50 years, little inferior to many other kinds, and arrive at length to a great size. Among the oaks of Hamilton-park (the property of the Duke of that name), so famous down to the end of the seventeenth century, there were trees which measured 27 feet round the trunk, with a wide expanse of branches.

The Horse Chesnut tree thrives well on the lower grounds, and deep soil only. The sweet chesnut, which quickly becomes a timber tree in districts more northern, does not succeed here. Its seasons of growth are too early, or too late, for the climate. In its first, it bears some resemblance to the Siberian pine, &c.; in its last, to the oak; its early growths being almost as early as the former, and its latter being nearly as late as those of the latter, and still more soft and susceptible of the cold. Hence its shoots are alternately put forth and destroyed, and it generally becomes a low stunted shrub. But this does not seem to have always been the case. There was in the lower part of the parish of Cambusnethan, on the grounds of Mr. Lockhart, of Castlehill, a tree of this kind, magnificently branched, and of a very large size, the stump of which still remains living. This seems to be a presumption of the truth of the conjecture advanced in the former part of this Report, that the natural accretion of flow mosses tended to increase the inclemency of the neighbouring climate; since the annual addition
which

which these mosses visibly acquire, in a district where they abound so much, may, in the course of two or three centuries, come to such an amount as to produce a sensible change in the state of the air; while in districts where the extent of moss is inconsiderable, such effects do not take place. The fate of the walnut, which may be considered as much a timber tree as a fruit one, is nearly the same with that of the sweet chesnut.

The Poplar delights most in water-formed soils, but is averse to marsh, and, when happily situated, makes quicker progress than any other tree. This county has been long in possession of two kinds of the white poplar, equally successful. Distinguishing them by their habits, they may be called the towering and the branching. The Lombard poplar, which, it has been said, becomes a large tree besouth the Trent, makes here but poor advances to timber. The balsam poplar makes very great progress.

In general, all the soils which lie immediately over the freestone rock, are much disposed to produce wood, and almost all kinds of trees thrive in them. Land lying on a quick declivity, where the water issuing from the veins of the earth flows freely away, is very favourable to the growth of wood. For this reason, trees grow better on the dip than on the crop of the mineral strata. Trees, planted by the winds, or by the birds, seem to thrive better than those cultivated by man. Whenever they meet on the same ground, the advantage which the child of Nature has over the child of Art, is conspicuous. There is a kind of coarse moorish soil, generally lying upon a thin bed of watery gravel, with an impermeable bottom under it, on which no trees will thrive till its faults are corrected. While the water below starves the roots, the close accumulation of heath and mosses on the surface, deprives them of the benign influence of the sun and air, and they pine and die in a few years. To raise a plantation, in such circumstances, it is necessary not only to give the ground a slight draining, but to destroy the obscene growth upon its surface.

In concluding this part of the subject, it is proper to observe, that though woods succeed, every other circumstance being similar, better or worse in proportion to the elevation in the atmosphere in which they are placed, yet they succeed worse on the summits in low elevations, than they do in much higher situations, where there are still greater heights around. The reason of this is too obvious to require explanation.

CHAP. VIII.

WASTES, WITH THE POSSIBLE IMPROVEMENTS OF THEM.

THERE is no land in this county over which the right of property is not claimed; but there are some high moorish grounds, not reckoned capable of any considerable improvement, on which the adjoining proprietors have a right of pasturage, according to some established proportion. If we call those lands wastes, which, from the high elevation in which they lie, the poverty of their soil,

soil, the ruggedness of their surface, &c. have been considered as no farther beneficial to man, than on account
of the sustenance which domestic animals can collect
from them, the extent of waste land is very considerable,
nearly the half of the surface. But the wastes which are
more peculiarly the subject of a work of this nature,
are those enormous masses of peat earth lying in a more
moderate elevation, and already described. The extent
of these dismal fields, their uselessness, and the probability of their injurious effects on the country around, have
been stated, and a despair expressed of any extensive
reformation being made upon them. But what is beyond the powers of man, when properly exerted! One
of those rare geniuses, the exertions of whose talents are
extensively beneficial to society, has appeared in the
adjoining county of Ayr, and put the art of reclaiming flow mosses into a train which may be carried to
a vast extent, and prove an universal benefit to the nation. If, according to the sentence which a witty author
of the last age puts into the mouth of his fabulous king
of Brobdignag, " He who has made two ears of corn,
" or two blades of grass, to grow on the spot which
" produced only one before, is a greater man than all
" the politicians of the universe*;" what gratitude
does the public owe to the man who has taught how
many ears of corn, and how many blades of grass, may be
produced on those wide wastes producing none before?
A particular account of the improvements on mosses
comes more properly into the Agricultural Report of the
county where they were first begun; and indeed the necessity of such an account is superseded, by a letter already in the possession of the public, from a gentleman

---

* GULLIVER's Travels.

of

of acknowledged ability, who has taken great pains to investigate and to state the different processes, and the result. But as every thing relating to a matter of such national importance ought to be recorded, it may not be amiss here, lest it should be omitted otherwise, to give some account of its origin. At the time of the American war, JOHN SMITH, Esq. of Swinrigmoor, in the parish of Dalry, then a youth, stung with the desire of military glory, left his property under the management of administrators, and went to gather laurels on the plains of America, leaving strong recommendations with his managers and tenants to cultivate some pieces of moss, hard by his house, by such means as were then known, in order to take away the unsightly appearance. The peace of 1783 put a stop to his military career, and he returned home to look after his private affairs, where he found that his recommendations respecting the moss had been little regarded. On one corner, however, which had been dug over, some lime in powder had been carelessly thrown, and some oats strewed. Here he observed, that wherever the lime had come, the oats sprung vigorously. Improving the hint thus given, he applied lime to moss, in various ways, and in different quantities, and by repeated experiments, found that a large dose of hot lime, applied to the wet surface of moss ground recently dug up, decomposed the parts with which it came in contact, and rendered a substance, formerly inert, highly fertile. He afterwards found, that raising potatatoes the first year after liming, still increased this fertility, by the stagnation of the air under the cover of this broad leafed plant: but to raise abundance of potatoes, it was found necessary to give the moss a small quantity of dung also. The ability and diligence with which Mr. SMITH has conducted these experiments, his steady perseverance, in

spite

spite of the obloquy and ridicule of a prejudiced neighbourhood, and the happy issue to which he has brought
these discoveries, do him great honour, and ought to
place him high in the list of those valuable characters,
by whose useful labours mankind have been benefited.

The operations performed in reclaiming moss, the expense attending them, and the general return obtained,
are all accurately stated in the above-mentioned letter,
to which those who are about to improve mosses are
referred. These operations are now begun in different
parts of this county, and are conducted mostly by people
not otherwise directly engaged in agriculture. It is
probable, the novelty of the thing, the prospect of immediate advantage, and of rendering the ground more
valuable in future, may induce others to purchase moss
grounds, and follow the same example, so as to direct
the employment of more capital towards the improvement
of the country. A few loose hints, which occurred on a
late survey of the reclaimed mosses in Ayrshire, shall,
therefore, be here submitted.

The altitude of these mosses is not considerable, perhaps not more than 150, or at the most 200 feet above
the level of the sea; but it seems evident, that moss may
be cultivated for corn to any altitude in which it succeeds
on the neighbouring fields; the corn on the mosses,
though much more luxuriant, being nearly as forward as
that on the hard ground around them.

Very slight draining appears to be sufficient in the first
preparation, and levelling is not attended to. The crops
were thriving where water was standing within ten or
twelve inches of the surface, and equally well on the
hillocks and little hollows. The draining can be pursued
gradually, as occasion requires; and levelling can be performed with much more ease in the future workings, as

L 3
the

the parts of the moss are more separated by the effects of the lime, labour, &c. The rough manner in which the moss is first turned over, seems not only to be sufficient, but preferable to a more accurate execution, as more surface is exposed to the action of the lime, and of the weather, during the winter. For these reasons, it is evident, that, if labourers were plenty, the first preparation of moss might be executed at a very moderate expense*.

Lime

---

* But the Ayrshire mode of cultivating peat seems to have proceeded on the popular notion, that it is a kind of earth, containing much inert vegetable matter, which the quick-lime was to decompose; but, in fact, it is altogether vegetable matter, into which mineral water has generally washed combinations of iron, and perhaps minute quantities of other compounds, and yields less earth in combustion than most vegetables do. A pound of peat, burnt on a clean iron plate, yields scarcely five drops of ashes, of which there is about a fourth of heavy charcoal, which, when washed, and put in the fire, burns with a blue flame, about a fourth of ochre, a very minute quantity of Epsom salt, about three times as much sea salt, gypsum about double the quantity of the other salts, some particles of shining sand, and the rest is lime. The great fault of peat, therefore, as a subject of cultivation for the production of useful vegetables, is the almost total absence of the earths of which soils are composed. The application of lime gives one species of earth; but though lime is an important ingredient, lime alone can never make a fertile soil : hence the addition of clay and sand seems to be necessary. Now clay and sand can always be got at an easier rate than burnt lime, as the nearest rising ground will always furnish those ingredients. The quality is a matter of indifference, since the most barren will render moss fertile as well as the richest. About 30 years ago, a few cart-loads of hard barren till, dug from the bottom of a pit more than 20 feet deep, was laid, by way of experiment, on the corner of a moss of great depth, and very soft and spongy. Where the hard substance was spread, the natural coarse herbage died away, and sweet grasses appeared in its stead. Part of this moss, including the spot where the till was spread, has been laboured, dressed with lime, &c. and cropped these three years; and the crops on this little spot have always been greatly more vigorous than those on the rest. This induced the farmer

Lime being the agent by which moss is converted from an inert to a fertile state, this species of agriculture can only

farmer to make a similar experiment on two contiguous pieces of peat ground. He spread the same kind of hard till on the one, and quicklime upon the other. The crop where the hard till had been laid was the best; and when the ground of both was again turned up, the contexture of the peat was more completely destroyed, and converted into a tender mould, than where the lime had been laid. From these facts, and many others which might be adduced, it is evident that all peat ground might be made productive of useful vegetables, by spreading not less than 100 cubic yards of heavy earth upon its surface; for notwithstanding the natural incorruptibility of peat, while it remains in its original conformation, it has been proved by many experiments, that whatsoever separates its parts, and destroys its natural contexture, enables it to yield food to growing plants. Now the clay and sand of which till is composed, by blending with the peat, not only effects this, but gives it somewhat of the qualification of a soil. Nor need the upper grounds be injured by the operations of reclaiming moss. Breaking in on the face of the nearest rising ground, the surface soil might be left, and the subsoil carried off, observing only to leave the bottom regular, and not go to a depth to retain standing water. The upper soil left might also be thickened and ameliorated, by bringing back the peat thrown out of drains to mix with it. But though the admixture of sand and clay will fertilize peat, lime is an important ingredient, which ought not to be omitted when it can be conveniently had. The farmer, whose experiment was just now stated, gave another dose of effet lime, the following year, to the piece which had been limed before, and his crop was better than that on which the till had been laid. Besides the effect which lime, when quick, has in decomposing the peat, all soils are more fertile by the admixture of calcareous earth.

When lime is at too great a distance, the best succedaneum for it is burnt clay. Every husbandman must have observed, that this has the same tendency as calcareous earth to promote the growth of sweet herbage; and corn has been found to tiller more abundantly when planted on peat mixed with brick-dust, than on peat mixed with lime. It might be prepared by burning any kind of cohesive earth in kilns, by means of peat fuel; or after the earth had been applied in a crude state, and thoroughly blended with the peat, by labouring and cropping for a few years, the ground might be turned over in the early part of summer, and

L 4

after

only be advantageously pursued where lime can be procured at a moderate rate. The noble Lord, whose useful
researches have been mentioned, was convinced that moss
could be rendered fertile by the application of alkaline
salts; and he thought these could be extracted from sea
water, so cheap and so plenty, as to be used for this purpose. Nothing of this kind, however, has yet appeared.
The seat of the moss improvements in Ayrshire, is peculiarly happy, beds both of coal and lime approaching
to the surface in different places. Lime is, therefore,
administered in large quantities, and the success justifies
the practice. But there is frequently much moss to re-
claim, in places where lime is less abundant; and it would
be of great importance to know if the purpose could be
effected by a more sparing application.

Different spots were shown on which the lime had been
applied in different states; and on those spots where it
had been applied, when recently slacked and still hot and
powdery, its effects were by far the most considerable.
Hence it would appear, that the causticity of lime, which

---

after being exposed to the drought for some days, set on fire. The
combustion of the inflammable substance would reduce the earthy part
to brick-dust, which, being mixed with the unburnt peat, would produce most luxuriant crops.

Dung has also been successfully applied in fertilizing peat. It has
been already observed, in remarking on Lord MEADOWBANK's compost, that fermenting dung, by changing the original conformation of
peat, renders it capable of yielding food to vegetables; but it appears
not to be a frugal practice to use the animal and vegetable substances of
dung, for which the husbandman has always much occasion other ways,
as an application to ameliorate vegetable peat, since the same thing may
be effected by less valuable substances; besides, the body of peat will
sooner be consolidated, so as to support the weight of labouring animals, and arrive at the true qualities of a soil, by the application of the
heavy earths, than by that of dung.

consumes

consumes the wet vegetable substances with which it comes in immediate contact, is its most important, though, perhaps, not its only effect. Might not, then, a less quantity, more accurately spread, as the dose was more sparing, and the causticity heightened by slackening it with boiling water (which in a coal country, might sometimes be done with little inconveniency or expense), decompose enough of the moss to make it fertile? In Ayrshire, after administering lime largely, a crop of potatoes, and four or five crops of oats, are taken without intermission or recruit. Where lime is scarcer, it might be frugal to administer it in small quantities, and more frequently; for example, a part for the first and a part for the third crop: and in this manner, new surfaces being exposed by every turning, it may be presumed that a less quantity of lime would decompose more moss.

Where the lime had not been applied till it had become cool and damp, the crops, though much inferior, were incomparably better than could have been produced without some such application : and, consequently, it may be presumed, that lime, besides its corrosive quality, possesses others tending, in some degree, to fertilize moss. And this, perhaps, is accomplished, not only by the calcareous substance, but by the effect which the particles of sand and clay, accompanying lime, have in consolidating the surface. To make up for the scarcity of lime, therefore, the schistus, mentioned under the head of manure, might frequently be applied along with it. Proofs of the good effects of this may be seen in the parish of East Monkland.

Besides fertilizing moss by means of lime, the surface, after being in some measure reduced to mould by the operations, might be carried as a manure to the solid grounds around, with great advantage; and thus the

fertility

fertility of all the fields in the neighbourhood might be increased, in a very high degree, without impairing that of the mosses, which having always plenty of depth, would be as fit for the action of lime as ever; and the surface of one acre of moss would be more than sufficient manure for six acres of firm ground.

But mosses may be profitable for grass, as well as for corn fields. It is represented in the letter above-mentioned, that after the culture of corn is abandoned, on account of the superabundance of esculent grasses which spring up amongst it, by the addition of rye-grass seed along with the last corn crop, a plentiful crop of hay, for the first year, is produced, and the pasture for succeeding years is worth 25s. per acre. The truth of this statement is not at all questioned; but it is evident, that none of the fields of moss, which have been cultivated for corn, and are now in grass, are brought to the best possible state for grass grounds; and there is a probability, that a little more attention to the culture of grass might be accompanied with success. It is probable, also, that a close cover of esculent grass would greatly facilitate the operation of any new application of lime, to increase still farther the fertility of mossy ground; the residue of such herbage on the surface, and its living roots spread through the soil, being much more susceptible of the putrid fermentation, than the dead plants of which the soil is originally composed.

Although complete draining at the first, while the great clods still adhere, and might wither too much in the summer's drought freely admitted into the interstices between them, be neither necessary or proper, it should certainly be pursued as the decomposition of the mossy substance takes place, that by the time the surface becomes fit to retain a sufficient quantity of the falling moisture,

moisture, it may be freed from the injury of the surplus stagnating in it. There seems to be a want of this complete draining in the late improved mosses which are now in grass. The turf, though thick in some places, is not regular; nor does the herbage appear to be very palatable to cattle. More accurate draining, and condensing the surface by repeated rollings, and spreading argillaceous substances upon it, would certainly improve the turf. If the drains were covered ones, and the surface well levelled, wherever springs on higher grounds were at command, there is, perhaps, no case in which watering, so much recommended of late, could be more easily and successfully practised.

But it would be necessary that these grounds were also replenished with the most proper grasses. The grasses of the culmiferous kinds, observed on them, were chiefly the grass called *poa purpurea*, and the soft millet grass *(holcus lanatus)*. Neither of these are much relished by cattle, or believed to be of the most nourishing quality, though eaten occasionally. The foxtail *(alopecurus arvensis et pratensis)*, the sweet scented early meadow grass, *(anthoxanthum)*, crested dogs-tail grass *(cynosurus crestatics)* and the *poa pratensis et trivialis*, would make a valuable addition. They are our best native grasses, and being all frequent, there would be no difficulty of gathering their seeds, and propagating them to any necessary extent. White clover appears in some places, which indicates that it may be farther propagated by sowing its seeds. The daisy, silverweed, crowfoot, &c. make up the list of the moss herbage, the place of all which would be much better supplied by the cow grass of the English farmers *(trifolium pratense)*, the branching of which enables it to take a fast hold of the soil. It is believed, that improved moss thus drained, condensed, and stored with

'with proper herbage, would be inferior to no other soil in the production of grass.

There is much mossy ground in this county, of too great elevation to encourage the cultivation of it for corn; yet its situation and circumstances are such, that, probably, lime might be applied to it with great advantage. It will be sufficient to give one instance, which will apply in all similar cases. Through the parishes of Lesmahagow, Douglas, and Crawfordjohn, there is a tract containing vast beds of excellent lime-stone. Among the wild sheep pastures of the same country, there are large fields, the general surface of which is tolerably regular, and the acclivities very moderate, and at the same time the bottom not very soft or miry; but being always drenched with the stagnant water retained on the surface, bear only herbage of the coarsest quality, chiefly stool-bent *(juncus squarosus)*, furnishing neither plentiful nor nourishing pasture. In the wide sheep pastures on the S. E. corner of Scotland, and those of England immediately bordering with it, where all the branches of the business of shepherdism are well understood, and diligently pursued, the active and intelligent occupants of sheep farms have, for some time, been in the custom of drawing small drains through all the moist parts of their pastures, to lead away the superfluous water which arises from springs, or descends from the hills in rains; and the general opinion is, that great benefit is derived from this practice. Were the same followed with the fields in question, and hot lime spread upon them, the coarse herbage would be consumed, and white clover and other sweet grasses soon appear in its room, which would probably render one acre of this pasture more valuable than six in its present state.

CHAP. IX.

LIVE STOCK.

SECT. I.—CATTLE.

THE number of oxen kept in this county is inconsiderable. Exclusive of those which are casually brought in, to fatten on the summer's pastures, or on turnip, the whole, perhaps, does not exceed 200. Milch cows, and young females rearing to supply them, are the principal stock. The whole may amount nearly to 30,000.

The neat cattle of this county are a mixture of many breeds, very different in figure, size, and proportions; many of them, perhaps, very ill adapted to the nature of the country. So far as attention has been paid to breeding, milk rather than beef seems to have been the object; and this object, perhaps, has not always been pursued with the greatest judgment. There are, however, exceptions to this stricture. Experience has shown, that cows of a bulky carcass are fit only for rich pasture, firm ground, and a sheltered situation. Those of a smaller size pass more easily over a soft soil, are more active in collecting their food on meagre pastures; and, as they require proportionally less food, have as many teeth, and jaws fully more nimble, have less trouble in ruminating the necessary quantity of dry fodder, and keep themselves in good habit at all times. Hence it is,

perhaps,

perhaps, that small cows, though they give less milk at a time, generally give it of a richer quality, and for a longer continuance, than large ones. Upon these principles, handsome cows, weighing from three to four hundred weight the four quarters, when fat, are bred in different parts of the county; and more attention has been paid, of late, to obtain the desired appearance. The colour is mostly brown, with spots of white, the hair thick set, soft, and sleek, the head and neck lean and slender, the ears small and neat, the limbs short, small and clean boned, the chest rather round than deep at the heart, the shoulders, and more especially the loins, broad and square, the back, from the shoulder to the descent of the rump, quite straight, the tail long and small. Some aim at having cows without horns; but when there are horns, they are small at the root, not long, and pretty erect.

This valuable breed of cattle are in greater perfection in the northern district of Ayrshire, and the neighbouring county of Renfrew; and it is probably from thence they are derived; numbers of young cows, from these quarters, being brought for sale to the fairs of Rutherglen. The inferior breeds of cattle are gradually giving way to these through the whole county; and the size of the cattle is increasing in proportion to their feeding.

SECT. II.—SHEEP.

In all the lower parts of the county, enclosing has, in a great measure, banished sheep; and that kind, of which the little flocks on the low ground was formerly composed, is now lost. Whether this is a disadvantage

or

or not, it is impossible now to determine. Where enclosures are made defensible, it is not uncommon to feed
sheep, bought in from breeding districts. These are
either ewes and lambs, or wedders, mostly the former;
but it is difficult to make a computation of the numbers
which are so fattened annually. A few tame sheep are
still kept on some of the low grounds, mostly mixtures
of the Dishley breed, less or more degenerated. The
black-faced mountain sheep are generally so restless and
impatient of restraint, that they can seldom be kept in
enclosures without injuring the fences, which has induced some to procure flocks of the quiet Dishley breed.
But the business of sheep pasturing is chiefly exercised in
the wild and mountainous parts. The sheep are of that
kind, distinguished by shepherds under the name of the
short moor sheep. They are so well known, that any
attempt to describe them would be superfluous. Long
experience has shown that this animal is well adapted to
the situation in which it is found: but it has been much
regretted, that so little benefit is derived from the fleece
as a material of manufacture. Great hope was entertained, that the patriotic exertions of the British Wool
Society would have made considerable improvements in
this respect. But a number of obstacles stand in the
way, which cannot be easily surmounted. A few of these,
set forth by some of the most intelligent sheep farmers
in this county, shall here be stated. 1st, The decided
preference which the Yorkshire jobbers (the principal
purchasers of surplus stock) give to short sheep, choosing
even the roughest woolled, and buying them at a higher
price, than those of equal size with a finer fleece. Thus
what is lost in the value of the wool is gained in the sale
of the stock; and shepherds are tempted to degrade rather than ameliorate the wool, by introducing rams of
the

the coarsest fleece to the breeding ewes. 2*dly*, The demand for coarse wool is greater than for fine; and when
the market is dull, the sale of the former is readiest.
From this state of the wool market, it is more the interest of sheep farmers to increase the quantity, than to
improve the quality of wool. 3*dly*, Sheep always thrive
best on the ground on which they have been bred perhaps for some generations; and, therefore, it is thought
imprudent to dismiss a known flock, and bring in a new
one; experience having shown, that immediate disad-
vantages frequently attend such a step; whereas the advantages are more uncertain and remote.

This useful animal is subject to various diseases, the
most fatal of which are the rot and the braxy. Scarcely
any effectual remedy has been found for either. A very
intelligent sheep farmer strews the branches of the Scots
pine on the pastures of the sheep of the first year (to
which the latter disease is chiefly incident), upon which
they browse; and he has found this a considerable pre-
ventive of the disease. If the regular study of the vete-
rinary art were more common, the practice of medicine
among domestic animals, whose manner of life is more
simple and natural than that of man, might perhaps be
more easy and successful. At least it seems worthy of
public attention, to make fair experiments, in order to
discover if any better means can be found for saving the
lives of this and other useful animals, than the quack
nostrums of ignorant and superstitious people.

The number of the standing stock of sheep in the
whole county is about 120,000.

SECT.

SECT. III.—HORSES, THEIR USE IN HUSBANDRY
COMPARED TO OXEN.

THERE are a great number of horses, of the draught
kind, bred in this county. The number employed in
agriculture, with the young ones rearing on the different
farms, amount altogether to about 8000. The number
of those kept for pleasure, for travelling, for the carriage of goods, &c. is not ascertained, but is certainly
very considerable.

The draught horses of Clydesdale have long been in
high estimation, and are so well known, that a description of them would be unnecessary. Dealers from different parts of England come to the Glasgow and Rutherglen markets to purchase them, and prefer them to
the Derbyshire blacks. Those of the upper ward, where
the greatest number are bred, are esteemed the best.
They have been sold, of late, in the Lanark and Carnwath markets, at three years old, from 20*l.* to 30*l.* and
upwards. And of late colts in the second year of their
age have sold at those prices. The older proportionally
higher. They have been much improved of late, and are
still improving, especially in size and weight.

Formerly oxen were used in tillage, in different parts
of this county; but when the progress of civilization
demanded better roads, and better roads were, of course,
obtained, husbandmen began to make more frequent use
of carriages, and to greater distances. From the first origin
of carriages in the county, there seems to have been a
predilection for the single-horse cart, the propriety of
which has been justified by more experience. In consequence of this, the horse, which was not only the most
ductile and expeditious animal, but whose hoofs were

CLYDESDALE.]   M   best

best adapted to receive an armature fit to defend them against the injuries of rough roads, obtained a general preference. The farms being small, one set of animals for home, and another for distant work, could not be kept; and thus the ox, being the least generally useful, has been gradually dropped. The use of oxen in the plough is not yet entirely abandoned. Along the eastern border of the county, they are still employed in that draught. But as a pair of oxen are joined in the same plough with four horses, to do what the horses might very well perform without them, the oxen seem to be little better than an useless incumbrance. What a pity, that those useful animals should be either abandoned, or so unprofitably employed !

Some gentlemen have again begun to use oxen for all the purposes of draught. The Right Honourable Lord DOUGLAS always works a few; and, at his Lordship's desire, his manager communicated the following comparative statement of the importance of the horse and the ox in labour, at the time the former Report was printed; and though the money value of both animals is different from what it then was, it shall here be stated in the same terms.

An ox at the price of 7l. 10s. is equally strong in draught with a horse at 20l., and equally fit for the plough, cart, or harrow.

The ox requires one-fourth less fodder than the horse, and only a little unthrashed oats, from an eighth to a sixth of what is requisite to support the horse ; and if 14 pounds of raw potatoes be given to the ox in a day, he will need no oats, and not consume more than half the fodder ate by the horse.

The ox may be wrought from four to ten years of age, and still increase in size, and be capable of carrying

ing more flesh, when he is turned to fatten; whereas the horse, in that time will lose one-fifth of his price.

The ox may be turned to pasture in summer, as soon as he is taken from the yoke; and will gather his own food, without needing any corn or attendance.

The ox is as much fatigued with seven hours work in the day, as the horse is with eight; and the execution of the ox is scarcely more than four-fifths of that of the horse, in the same time.

After the ox has filled his belly, he must have time to ruminate, and therefore cannot be baited, and put to work a second time the same day, like the horse, without being greatly injured.

From this view of the matter, it seems evident, that oxen could not be advantageously put to all the purposes of labour now done by horses; but, at the same time, there might be a great saving made by using them in part. To give one short example, for the sake of illustration: On a farm where there was frequent occasion to plough stubborn land with a deep furrow, a pair of oxen might be very properly joined to a pair of horses in one plough, for that purpose; when less strength of draught was requisite, the horses and oxen might be wrought separately; or in any case where distant carriages were necessary, when a longer continuance at work was required than was suitable to the nature of oxen, or the state of the roads unfavourable for their feet, the horses might be employed in such, while the oxen were forwarding the work at home. The saving which would accrue from the difference of the first cost of oxen and horses, the difference of the expense of maintenance, &c. of the two animals, avoiding the fall of price by the sale of horses, and the total loss of them by incurable lameness, &c. is left to the calculation of the reader, and will ap-

M 2

pear

pear so considerable, that it is probable the partial use of oxen needs only to be fairly introduced, to become general: and thus the ox would be restored to his real importance, and be found no less valuable as a labouring than as a feeding animal.

SECT. IV.——INFERIOR STOCK.

BESIDES the kinds of live stock above enumerated, scarcely any other can be said to be an object of attention among the husbandmen of this county. A kind of Jewish abhorrence of swine seems to have taken place, about the rigid times of the Reformation, in the western counties of Scotland. They were unclean beasts,—it was sinful to eat their flesh,—and neither creditable nor profitable to keep them. And though these prejudices are now pretty much worn out, pork is not yet, in general, a favourite food, and, of course, the number of hogs kept and fed are not considerable. Country gentlemen frequently keep a small piggery to serve their own tables; and some few pigs are bought in by farmers, in different parts of the county, to consume the whey of the dairy in summer, and are fed upon potatoes, &c. in autumn. To serve this demand, some people find their account in keeping brood swine. But whey is now much used in printfields, as an acid, and the farmers in the neighbourhood of Glasgow sell their whey there for that purpose *.

There is no rabbit-warren in the county, but one belonging to His Grace the Duke of HAMILTON; and from a

---

* Since this was written, the mineral and vegetable acids have taken the place of whey.

single

single instance in a great tract of country, and subject to the inroads of every kind of depredatory vermin bred in the neighbourhood, no fair inference of the value of warrens can be drawn. But it may be said, that rabbit-warrens, and the cultivation of a country for corn, seem to be somewhat incompatible.

Geese and turkeys are bred mostly by people of fortune, for the sake of variety at their tables.

Dunghill fowls, and sometimes ducks, are found in all the farm-yards of the county. The former, particularly, are equally inimical to the kitchen garden and the corn field, both in spring and autumn: and for that reason are commonly the detestation of husbandmen, who think the injuries they do at these times are greater than all the value they yield through the year. The housewife, however, reckons most on the profit and conveniency of fowls and eggs, which of late have sold very high; and under this favour they are sheltered and preserved. A fowl has a great appetite, and if it received all its food from the hand, would soon consume its own value; but as they pick up much of what would otherwise be lost, and make an agreeable variety at the table, they are a conveniency, the want of which would be much felt. But the profit arising from keeping them cannot, upon the whole, be considerable.

Pigeons are considered by husbandmen as a nuisance, and the laws by which pigeons and pigeon-houses are protected, giving one class of men the privilege to let loose a flight of animals to prey upon the property of another, as grievous and degrading. The injuries done by pigeons are very considerable, but they last only for one month in the year; that is, from the time the corn has begun to ripen till the harvest home. One would think it reasonable, that the same laws which bestowed

M 3

this

this privilege, should have obliged those on whom it was bestowed, to provide plentifully for their pigeons during that month. If this were done, it would not be difficult to scare them from the corn fields. At all other times pigeons are not very hurtful. In seed time, as they neither turn up with bills, nor scratch with their claws, they devour no part of the seed, but that which has been left uncovered. In winter, their craws are found crammed with the seeds of weeds which infest corn fields; and, in this instance, they may be presumed to be of some service to husbandmen: but it appears to be very inconsiderable indeed; for the fields to which they have oftenest resorted, in winter, have still too many weeds next summer. Rooks are regarded by husbandmen pretty much in the same light as pigeons. In harvest they are not much less destructive, and in seed time much more so, as they turn up, with their strong bills, and devour the seed, both before and after it has begun to vegetate. The only compensation they make, is picking up some of the small earth vermin, after the plough. They are not protected by the same law as pigeons, but the proprietors of plantations where the rooks nestle, may, and sometimes do, prevent people from going into these plantations to destroy them. There is little reason, however, to complain of the proprietors of this county, many of whom having not only allowed, but encouraged the destruction of rooks. It is a matter which merits general attention, as the numerous plantations in the county afford room for so many new colonies of rooks, that they would increase in an alarming degree, if pains were not taken to destroy them.

Husbandmen are too much engaged otherwise to attend to the apiary, and of course there are not a great many hives of bees kept by people of that description.

There

There is no doubt but bees are profitable at times, when they succeed: but the greatest part of the information got from bee masters here, consisted of long accounts of their losses by the death of bees in bad years, and by the theft of hives in good ones.

CHAP. X.

RURAL ECONOMY.

THE title of this chapter, taken in its large sense, may be understood to comprehend the conduct of the husbandman in the course of his business, through all the varied seasons of the year. But as it has been the chief business of this work, after giving some description of the face of the county, to represent what was doing in the cultivation of it, it is presumed the subject is pretty much exhausted. However, such gleanings as remain shall be here collected; and these will easily comprise any miscellaneous topics relating to agriculture, and anticipate the purpose of a chapter of that kind. This chapter will therefore contain the following sections; viz.

1st, Agricultural Societies, showing the means followed by husbandmen for their mutual instruction and support.

2d, The

2*d*, The weights and measures used for ascertaining the quantum of things relating to a farm.

3*d*, The prices of the different commodities produced from farms.

4*th*, The labourers employed in agriculture,—hours of labour,—price, &c.

5*th*, Their accommodation, such as food and fuel.

SECT. I.—AGRICULTURAL SOCIETIES.

HUSBANDMEN, as a class of men, are less connected than any other of which the general mass of society is composed. There is no bond of union to conjoin them for the general interest of their order—no rallying point around which they can assemble for their mutual support: consequently, insulated husbandmen can never match in contest with the members of any other class of society. This assertion will be derided by superficial observers; but, it is presumed, will scarcely be refused by any person of candour, who soberly considers the matter. It may be attributed, in part, to their dispersed situation, and sequestered manner of life, but still more to the unjust prejudices which have long and strongly obtained among the other classes, and the laws and customs founded on these prejudices tending to depress and discourage them; and not to any contracted selfishness of disposition attached to the profession. On the contrary, we find no class of men so liberally disposed to assist one another in forwarding their mutual labours, or in relief of the emergencies which occur to an individual; and none so free in communicating such knowledge as they possess, for the benefit of their brethren.

For

For such purposes, several societies have been instituted in different parishes or other districts of this county. Some are of a considerable standing, and some more recently formed. The members of these meet at an agreed place, perhaps once a month, where they converse about the operations in agriculture, in which they have been employed, and the success attending them. A subject of discussion is also proposed at one meeting; and the members take it into consideration, and deliver their opinions at the subsequent one. There are instances of members of some of these societies attending the Lectures on Agriculture given by Dr. COVENTRY at Edinburgh, as well for the instruction of their society as for their own. They speak highly in praise of this gentleman's labours.

Some of these Agricultural Societies have still another object. A stock purse is formed by stated contributions paid by all the members, out of which some relief is given to persons connected with the society, who may fall into accidental distress.

A Corresponding Agricultural Society has been lately established at Glasgow, upon a large scale. The members already admitted amount to several hundreds, and are still increasing. Each member, upon his admission, contributes a guinea to the common stock. The Society, at their general meetings, are to elect certain annual office-bearers, and appoint a committee to assist them. These are to keep up a correspondence with the lesser Societies around, receive such communications as they may send, and return copies of such as are received from other quarters, &c. The stock is to be lent out on interest, only to husbandmen who are members of the Society.

SECT.

SECT. II.——WEIGHTS AND MEASURES.

The Dutch and Trone weight, the only kinds used in rural commerce, in this county, are described in the Mid-Lothian Report. The proportion they bear to avoirdupois weight, is the same in this as in that county. Meal only is sold by the former; butter, cheese, wool, flax, butchers' meat, and hay, by the latter.

Eight *lb*. make a peck, 16 pecks a boll of meal; 16*lb*. of butter, cheese, flax, wool, and hay, make a stone, 12 stone of wool a pack. Though there is a clause almost in all leases, to restrain the farmer from disposing or carrying off fodder from the farm, hay, being rather a new product, is not understood, and there is always a great deal of hay sold by farmers. Five stone of hay is nearly equal to a hundred weight, and consequently 100 stones near a ton. Though common farmers are not allowed to sell straw, there is always much straw bought and sold; and the most secure way of ascertaining its value, which is now frequently resorted to, is to weigh it as hay.

It would be by far the most just and accurate way of ascertaining the value of grain of all kinds, to weigh it, and should certainly be used in all cases.

Lineal, square, and liquid measure, are the same here as described in the Mid-Lothian Report.

In the dry measure, used in the sale of grain of all kinds, a boll contains four firlots, a firlot four pecks, and a peck four forpets or lippies; 16 bolls make a chalder. The firlot used to measure barley and oats, is almost one half larger than the firlot for measuring wheat, beans, pease, &c. Both these measures are about one-sixteenth larger than the Linlithgow standards of

the

the same denominations. But for more than 30 years past, wheat has been bought and sold by the Linlithgow standard, which is now attempted to be introduced for other grains.

In the lower parts of the county potatoes have been measured, for these 40 years, with a dish of the shape of a cask, the peck measure holding 15 Scots pints; its full of potatoes, recently dug, weighs 43 *lb.* avoirdupois. In the higher parts of the county potatoes are sold by the barley measure.

The peck, or sleek, for measuring pears and apples holds about 18 pints. The confusion occasioned by the irregularity of weights and measures, is too obvious to require any comment.

SECT. III.——PRICES OF THE COMMODITIES PRO-
DUCED FROM FARMS.

THE commodities derived from farms, are either the materials of manufacture, or provisions for man and subservient animals. Both these are either obtained directly by means of cultivation, or indirectly from the animals supported on farms. The money price of provisions, of the first kind, have not risen in proportion to the articles which enter into the cost of raising them; and indeed the price of the former seems scarcely to have been affected by the rise of the latter, but fluctuated only on account of temporary starts of scarcity or plenty. In the course of the last 40 years, land rent, the wages of labour, the price of labouring horses, &c. is nearly tripled, but the price of grain has seldom been much lower during that period, than in the beginning of the present summer,

summer, 1797. On the other hand, the last kind of provisions, which require less additional cost of labour, has been regularly upon the rise, and the price is tripled, and in some instances quadrupled, in the above period. The current prices of the materials of manufacture seem more to resemble those of the first than of the last kind of provisions.

The following is a Table of the prices of Farm Commodities, about the time of the last publication, compared with the prices of the present and last year.

	FORMER.	PRESENT.
Moorland wool per stone,	from 6 s. to 8 s.	from 8 s. to 10 s.
Flax per ditto,	10 s. 6 d. to 1 l.	1 l. to 1 l. 10 s.
Butter per ditto,	16 s. to 1 l.	1 l. to 1 l. 4 s.
New milk cheese per ditto,	8 s. to 9 s.	10 s. 6 d.
Skimmed milk dit. per ditto,	6 s. to 6 s. 6 d.	8 s.
Butcher meat of all kinds per pound,	7 d. to 10 d.	9 d. to 1 s.
Fowls per pair,	2 s. 6 d. to 3 s.	4 s. 6 d. to 5 s.
Eggs per dozen,	6 d. to 1 s.	7 d. to 1 s. 4 d.
Hay per stone,	5 d. to 8 d.	5 d. to 8 d.

Straw sells at from one-half to two-thirds the price of hay.

The following Table of Fiars, or medium prices, will give the best idea of the price of grain. It contains only the prices of oat-meal and barley. The price of a boll of wheat may be computed to be nearly equal to that of a boll and a half of oat-meal, but now exceeds the price of two bolls. There are four different offices in the county, at which proof of the prices are annually taken; viz. the Commissary's of Glasgow, the University's, the Commissary's of Hamilton and Campsy, and the Commissary's of Lanark.

TABLE;

TABLE,

Of the Fiars of the Commissariot of Hamilton and Campsy, for Oat-Meal and Barley, for the following 60 Years, in four Periods of 15 Years each.

1st Period	Meal. £ s. d.	Barley. £ s. d.	2d Period	Meal. £ s. d.	Barley. £ s. d.
1737	0 10 0	0 8 $11\frac{6}{12}$	1752	0 14 $5\frac{4}{12}$	0 13 $8\frac{3}{12}$
1738	0 10 0	0 9 $7\frac{10}{12}$	1753	0 13 8	0 13 2
1739	0 12 $6\frac{8}{12}$	0 12 9	1754	0 11 0	0 9 $6\frac{11}{12}$
1740	1 0 0	0 17 11	1755	0 13 $10\frac{8}{12}$	0 12 $1\frac{6}{12}$
1741	0 9 8	0 9 $8\frac{2}{12}$	1756	0 18 11	0 17 2
1742	0 8 4	0 8 $7\frac{2}{12}$	1757	0 15 0	0 13 $8\frac{9}{12}$
1743	0 7 0	0 7 $6\frac{6}{12}$	1758	0 10 0	0 8 $6\frac{8}{12}$
1744	0 10 5	0 9 11	1759	0 9 8	0 9 $8\frac{4}{12}$
1745	0 16 $2\frac{2}{12}$	0 15 $3\frac{3}{12}$	1760	0 10 4	0 9 $7\frac{7}{12}$
1746	0 11 0	0 11 $1\frac{9}{12}$	1761	0 12 0	0 10 $6\frac{8}{12}$
1747	0 9 0	0 9 5	1762	0 17 4	0 13 $11\frac{4}{12}$
1748	0 10 0	0 10 $8\frac{6}{12}$	1763	0 13 0	0 14 $4\frac{6}{12}$
1749	0 10 $6\frac{8}{12}$	0 9 $0\frac{4}{12}$	1764	0 15 4	0 13 $10\frac{4}{12}$
1750	0 11 8	0 10 $7\frac{4}{12}$	1765	0 18 0	0 18 $3\frac{4}{12}$
1751	0 15 $1\frac{6}{12}$	0 13 $4\frac{6}{12}$	1766	0 18 0	0 19 $9\frac{8}{12}$
	£.8 11 6	£.8 4 $6\frac{10}{12}$		£.10 10 7	£.9 18 $2\frac{3}{12}$
Aver. of 1st period	£.0 11 $5\frac{3}{12}$	£.0 10 $11\frac{7}{12}$	Aver. of 2d period	£.0 14 $0\frac{5}{12}$	£.0 13 $2\frac{6}{12}$

3d Period	Meal. £ s. d.	Barley. £ s. d.	4th Period	Meal. £ s. d.	Barley. £ s. d.
1767	0 16 0	0 15 $0\frac{6}{12}$	1782	1 1 4	1 1 0
1768	0 13 8	0 12 $2\frac{4}{12}$	1783	0 16 8	0 15 0
1769	0 15 $9\frac{4}{12}$	0 15 $4\frac{1}{12}$	1784	0 16 0	0 14 8
1770	0 15 8	0 15 $6\frac{5}{12}$	1785	0 13 8	0 11 9
1771	0 15 $10\frac{2}{12}$	0 16 4	1786	0 15 $3\frac{4}{12}$	0 11 4
1772	0 17 $9\frac{4}{12}$	0 17 $1\frac{1}{12}$	1787	0 15 10	0 12 1
1773	0 17 0	0 16 $11\frac{8}{12}$	1788	0 13 9	0 12 6
1774	0 15 4	0 16 $3\frac{6}{12}$	1789	0 14 6	0 14 $5\frac{6}{12}$
1775	0 13 0	0 13 $9\frac{6}{12}$	1790	0 16 6	0 12 8
1776	0 14 0	0 12 0	1791	0 16 0	0 15 0
1777	0 14 4	0 12 2	1792	0 17 5	0 17 $4\frac{5}{12}$
1778	0 14 8	0 12 4	1793	0 17 6	0 15 $4\frac{9}{12}$
1779	0 12 4	0 10 8	1794	0 17 $0\frac{4}{12}$	1 0 $9\frac{4}{12}$
1780	0 15 4	0 11 $11\frac{2}{16}$	1795	1 1 1	1 3 3
1781	0 14 0	0 12 $8\frac{9}{12}$	1796	0 16 7	0 18 $11\frac{4}{12}$
	£.11 4 $8\frac{10}{12}$	£.10 10 6		£.12 9 $1\frac{8}{12}$	£.11 16 $2\frac{5}{12}$
Aver. of 3d period	£.0 14 $11\frac{9}{12}$	£.0 14 $0\frac{4}{12}$	Aver. of 4th period	£.0 16 $7\frac{3}{12}$	£.0 15 $8\frac{10}{12}$

(in which there have been two years of remarkable dearth.)

To

To which we now annex the prices for the seven suc-ceeding years.

	Oatmeal per boll.			Barley per boll.		
	£.	s.	d.	£.	s.	d.
1797	0	15	11	0	14	4½
1798	0	17	6	0	16	6
1799	1	16	2	1	6	3
1800	2	5	4	2	9	8
1801	0	19	0	1	2	4
1802	0	19	0	0	13	7
1803	0	19	6	0	16	4
	£.8	12	5	£.7	19	0½
Average	1	4	7½	1	2	8½

N. B. The retail market price is always somewhat higher than the fiars, so that the poor, who buy only a peck of meal at a time, may pay nearly, at an average, through the year, three halfpence a peck more.

SECT. IV.—LABOURERS EMPLOYED IN AGRICUL-TURE.—HOURS OF LABOUR,—PRICE, &c.

It has been already said, that the greatest part of agricultural labour is performed by servants hired by the half year, and living in the farm-houses. In many parts of the county, the women-servants work along with the men, at almost all kinds of out-work. But as more hands than ordinary are needed for cutting down the corn in harvest, many husbandmen, to secure a fixed number for that purpose, when they can be got, con-tract with villagers to assist during the time of reaping.

All

All these labourers have no fixed hours, but continue their labours while light and weather admit, and circumstances require. The poor girls, when light is gone, and the men sat down by the fire, resume their household labours.

Labourers, both men and women, are sometimes hired in by the day, particularly in the times of planting, sowing, and hoeing turnips and potatoes, hay-making, and harvest, when a fixed number are not provided. These work only ten hours in the day, beginning at six in the morning, and stopping at six at night; and taking an hour to rest at breakfast, and another at dinner. Hired labourers in winter, take breakfast before they go out in the morning, make a short pause to eat a little at midday, and quit when light fails in the evening. All labourers, have improved in dexterity and execution, but have lost much of the conscientious anxiety to forward the work in hand, which formerly appeared; and having too much of the disposition of the hireling who longs for the going down of the sun, it is doubtful if they do more work than formerly. Indeed it is a common observation of employers, that the higher the wages the less work is done. Day labourers mostly provide their own food.

The wages of the different labourers employed on farms, both when this Report was last published and during the last twelve months, are stated in the following table:

	FORMER.	PRESENT.
Wages of men servants, besides bed board and washing, per annum,	from 13*l.* to 16*l.*	from 20*l.* to 25*l.*
Wages of a maid servant, besides bed board and washing, per annum,	—— 5*l.* to 7*l.*	—— 8*l.* to 9*l.* 10*s.*

Wages

	FORMER.	PRESENT.
Wages of a harvest man for the reaping season (besides food and bed), - -	1 *l.* 14 *s.* to 1 *l.* 16 *s.*	2 *l.* to 2 *l.* 5 *s.*
Wages of a harvest woman for the reaping season (besides food and bed), - -	1 *l.* 8 *s.* to 1 *l.* 10 *s.*	1 *l.* 12 *s.* to 1 *l.* 16 *s.*
Wages of a labouring man per day at ordinary work, without food, - -	1 *s.* 4 *d.* to 1 *s.* 6 *d.*	1 *s.* 10 *d.* to 2 *s.* 2 *d.*
Wages of a mower, - -	2 *s.* 6 *d.*	2 *s.* 6 *d.* to 3 *s.*
———— of a carpenter, - -	1 *s.* 8 *d.*	2 *s.* 6 *d.*
———— of a mason, - -	2 *s.* 6 *d.*	2 *s.* 8 *d.* to 3 *s.*
———— of a thatcher, - -	2 *s.*	2 *s.* 6 *d.*
———— of a woman at field labour,	10 *d.*	1 *s.*
———— of ditto in harvest, -	1 *s.* 6 *d.*	1 *s.* 8 *d.*

Work by the piece is too often deficiently executed, when opportunities of slighting offer. When this can be avoided, it is, no doubt, the best way for both parties. The prices differ so much with circumstances, that it would be difficult to describe them with accuracy. It is sufficient to say, they are in general a good deal higher than those mentioned in the Mid-Lothian Report. But, notwithstanding, all work, the execution of which could be safely intrusted in that way, was also cheaper done by the piece than by day labour, till within these few years, that the war, and other circumstances, having swept away the labourers, competition is destroyed, and an undertaker can always contend for his own price.

SECT. V.—THE ACCOMMODATION OF THE LA-
BOURERS IN AGRICULTURE, RESPECTING FOOD
AND FUEL.

OAT-MEAL, potatoes, and milk, either skimmed or butter-milk, are the principal component parts of labouring people's food in this county. The servants maintained in farm-houses, always breakfast on a kind of hasty pudding, made of oat-meal, well known through all Scotland, and some of the northern parts of England, by the name of porridge. This is eaten mostly with butter-milk, and is not only wholesome and nutritious, but very grateful to the stomach of all who have been habituated to it. The bread is either thin unleavened cakes of the same meal, baked on an iron plate hung over the kitchen fire, called the *girdle*, or bread made of pease or barley-meal prepared in the same manner. Farmers, for the use of their families, salt up beef in the month of November. A portion of this is boiled twice or thrice a week, so long as it lasts, in broth, in which husked barley, cabbage, greens, or other culinary vegetables of the season, are mixed. The broth and beef make the dinner, while they last; and cheese, and sometimes butter, &c. are served in the intermediate days. Potatoes are often used in place of bread; and they, or porridge, with milk, make the supper. Butcher meat is less frequent in summer, when milk is plenty, but is always provided in harvest. Herrings, sometimes fresh, but more frequently salted, make also a part of the provisions.

It may well be supposed that day labourers, who have not only themselves, but generally families to provide for, are less plentifully fed than farmers' servants; but their manner of living, so far as they can attain, is pretty

CLYDESDALE.]　　　　　N　　　　　much

much the same. Very little butcher meat is consumed
in their families. Herrings and potatoes make a frequent
meal. When circumstances admit, those who can afford
it keep a cow, sell part of the milk, and consume a part
at home. Tea and wheat bread are the prevailing
luxuries among the lower ranks. Few are the luxuries
of the poor ! and it would be cruel to grudge them such
as they can attain. Yet it is to be regretted that the use
of wheat bread is become so general among them. The
Scottish nation has long found a wholesome support in
oat-meal. Oats require less cultivation, and can be raised
in much greater quantity than wheat ; and consequently
the supply must be more liberal, when the former, rather
than the latter, is the chief basis of food.

It has been already shown, that coal is abundant through
a great part of the county. The material of peat is not
less so ; and peats were formerly made by all the inhabitants residing near the mosses : but they are now too
much engaged otherwise to spend their time in making
peats. And though the price of coals, including carriage,
is tripled within these 40 years, they are found the
cheapest, as well as the best fuel ; and the use of peats is
almost abandoned, except in the upper parts of the county,
where the coal is most distant.

CHAP. XI.

POLITICAL ECONOMY, AS CONNECTED WITH OR AFFECTING AGRICULTURE.

———

SECT. 1.—ROADS *.

THE first Scots statute which provides for the making and upholding of public roads, is that of CHARLES II. anno 1669, cap. 16. This statute ordains, That all tenants and cottars shall be called out on the highways, with all their carts, sledges, spades, shovels, picks, mattocks, &c. to work six days in the year between bear seed time and harvest, for three years, and four days in the year ever afterwards. To make up the deficiency which this strength might not be able to effect, heritors are authorized to assess themselves to a certain extent, in proportion to their property, to be laid out in building and repairing bridges, &c. This law, which has been somewhat modulated and altered by after statutes, is the basis of the regulations for making and mending roads. It was no doubt the readiest expedient for the purpose ; and, perhaps, in the then circumstances of the times, there might have been difficulty in finding a

* In this section, written in autumn 1797, no material alterations seem to be requisite, as it will nearly serve, by altering the dates, for the present state of the roads in the county.

better.

a better. It savours strongly, however, of the barbarous notions of feudal times, when the most useful class of society were regarded as slaves, subservient to the pleasure of their superiors. And it seems to be equally impolitic and unjust, to abstract from its proper purpose, so large a portion of the labour destined for the cultivation of the country, and to lay on one particular class the burthen of making roads intended for the general benefit of the whole society. Husbandmen appear to have always regarded the statute work laws in this light, and thus contracted such an aversion to the duty imposed by them, that it has been, for the most part, very reluctantly and slovenly performed; and, accordingly, the roads were never in condition to answer the purposes of internal commerce. As that began to extend, therefore, some new expedient became necessary, and turnpike laws, which, in a more equitable way, lay the burthen of making and maintaining a road on passengers, in proportion to the use they make of it, were introduced. But this regulation by no means lightened the burthen formerly imposed on the cultivators of land. While the husbandman, like others, paid his tolls on the turnpike road, he was still liable in the old statute duty. In some instances, indeed, manure for land was exempted from tolls; and at length new acts were obtained, to convert the statute labour into money, and commutations were accepted, moderate when compared with the present value of the labour.

The making of roads is the first step towards the improvement of any country; and, in a country where the soil is naturally soft and retentive of water, no improvement can take place, till good roads of communication are made. Accordingly we find that the progress of agricultural improvement has uniformly followed the making

making of roads in this county. Turnpike roads were first introduced about the year 1755, in making the road between the cities of Edinburgh and Glasgow, by the Kirk of Shotts, and by Hamilton to Ayr; and though on account of the total ignorance and inexperience in the business, and the difficulties to encounter from the nature of the soil and materials, this was a very arduous unpromising undertaking at the beginning, yet by the laudable perseverance of the gentlemen of the county all difficulties were overcome, and the road has become of no less consequence in promoting improvements in agriculture, along its course, than in facilitating internal commerce. New houses were built; the fields were enclosed and subdivided; and, from the easier conveyance of manure, a new appearance of fertility given to the country. Since this first essay, the number of turnpike roads has been greatly multiplied. The same public spirited zeal has been exerted through all the county. Landholders, wealthy individuals, and bodies corporate, have united in obtaining Acts of Parliament, and advancing money for making roads through different tracks, in various directions; so that convenient communications are now opened, towards all quarters, through this county, to the most distant provinces of the kingdom; many of them passing through tracks which were almost entirely deprived of the benefit of intercourse before, there being either no roads, or such as were not passable for loaded carriages. Besides the building of bridges, and other appendages to the roads, the making of the roads themselves has, in some places, been very expensive. The soft sand-stone found in such places proving insufficient, the roads have been obliged to be made with hard stone, sometimes brought from rocks at several miles distant: hence the expenditure necessary for the accomplishment

of

of all these great works has been immense; and if time
and room would allow to state it here, would give a very
advantageous idea of the liberality of the gentlemen of
the county. In several instances, the revenue arising
from the tolls is very inadequate to the expense of making
the roads, not yielding perhaps above three per cent. of
interest on the outlay; but the creditors have the satisfaction to see, that they have contributed much to the
improvement of their country, and all of them are, in
some measure, partakers of the advantages arising from
it. Nor has the spirit for making roads been confined
to the great lines. These opening the communication,
branches of inferior note have every where been directed
towards them. The statute work of the different parishes
has been generally converted into money, and many who
have interest in particular roads, upon the faith of being
paid by instalments from these funds, have advanced the
expense, and made them by anticipation, so that the principal parish roads are mostly in pretty good order.

It has rather been unfortunate, that this county em-
barked so eagerly in road making, before all the requisites
of a good road were sufficiently understood. The old
roads, without regarding inequalities in the course, generally proceeded something near the direction which
seemed to be the readiest to some known station, and so
on from that to the next. Except a little financial
straighting, pretty much the same conduct was followed
in laying out the new. The inconveniency of steep pulls,
was lost in the main and more obvious aim of making
roads regular, smooth, and firm, instead of the former
awkward, rugged, and miry ones. From this, and other
circumstances unavoidable, where human fallibility is
the guide, and so many people of different views and
opinions concerned, mistakes have been committed, and
the

the roads have not at all times been conducted in the most eligible course. But where the main design was so public spirited—where so many sacrifices of private interest and personal ease have been made—and where the general result is so beneficial—it would show an unpardonable want of candour, to censure little errors with severity. Now that experience has shown the disadvantage of carrying a road over knolls, the pulls have been eased, and the early made roads much improved, wherever the funds would admit. In laying out the later ones, more enlarged ideas have taken place,—former errors have been avoided,—and instances of great judgment have been shown, in shunning the natural difficulties of the country, and conducting the road by the easiest route. This is particularly the case in the roads from Glasgow, by Muirkirk, to Dumfries, &c. and that which crosses it at Kilbride, leading from Hamilton to Ayrshire. The road now making, from Edinburgh, by Airdrie*, to Glasgow, is also conducted in a remarkably level easy line, as are some new roads near Leadhills. Upon the whole, notwithstanding the softness of the soil, the inequalities of the surface, and the great expense necessary for making and supporting roads to stand the fatigue of so many heavy carriages, the roads of this county are so many and so well kept, as to answer all the purposes of an extensive inland commerce. And perhaps no where in the kingdom are travellers better or cheaper driven, or commodities carried at an easier rate. Nor are there many instances, probably, of greater

* This road is now completed. It has been executed at an enormous expense; but to the honour of the gentlemen under whose auspices it has been conducted, it may be said, that it is in no respect inferior to any other road in the island, of the same length.

N 4 weights

weights being drawn; for it is not uncommon for a single horse to draw a ton and an half on a cart.

SECT. II.—CANALS.

THE Forth and Clyde navigation is so well known, that a minute description of it would be superfluous; and, besides, its course in this county is short. At 156 feet above the level of the sea, it enters the north corner of the county, in the parish of Cadder, and goes again out of it, into Dumbartonshire, at the aqueduct bridge over the Kelvin. This bridge has been much visited by strangers, and is admitted to be a stupendous work of its kind. The length of the canal's course within the county is eight miles. At a place called Stockingfield, a collateral branch is brought off towards the city of Glasgow. It ended at Hamilton-hill, an eminence at a little distance from the city, and was afterwards made to approach somewhat nigher. At the end of this branch a large bason is formed, and granaries, storehouses, &c. built around. This place is called Port Dundas, and stands on the top of a little steep hill within half a mile of the Cross of Glasgow. The length of the branch is two miles and three quarters.

There is another inland navigation in the county, called the Monkland Canal. It was begun by the authority of an Act of Parliament obtained in 1770, and carried from the centre of the coal works in the parishes of Monkland, upon a level of 252 above the sea, as near Glasgow as the same level would admit of its approach. The principal intention of this undertaking was, to increase

the

the sale at the Monkland collieries, and furnish Glasgow with a more plentiful supply of fuel. The coals were brought in boats to the extremity of the canal, and from thence let down an inclined plain formed of wood. At the foot of this they were again put in carts, and carried to the town. The circumstances of the times having occasioned the demand to be less considerable than was expected, and the repeated ladings and readings of the coal having inflamed the expense of the carriage, and injured the fuel in the opinion of consumers, the subscribers found the canal business an unprofitable concern, and came to the resolution, in the year 1782, to dispose of the stock by public auction. Mr. STIRLING of Drumpellier, and the mercantile company with which he was connected, became at last the sole proprietors of this navigation. An agreement having been made between the proprietors of the Forth and Clyde navigation and them, for their joint advantage, the canal has been carried eastward, to receive a supply of water from the North Calder, the source of which is a lake on the summit of the country, upon the confines of the county of Linlithgow, and westward to join the branch of the Forth and Clyde navigation at Port Dundas. The length of this navigation is about 13 miles. It is raised at the west end, from Port Dundas, by eight locks, 96 feet, and at the east end, near Airdrie, to bring it in a level with the channel of the Calder, by two locks, 21 feet. Its width at the surface is 30 feet, and 15 at the bottom; the depth of water about five feet. By this canal, coal, &c. is carried westward from the collieries, and dung, lime, &c. eastward, as manure for the adjacent lands.

A more extensive canal, to pass from Glasgow to Edinburgh through the middle of this county, was projected a good many years since: in the year 1792 there was a

great

great prospect of it being carried into execution. The principal intention of this undertaking was, to furnish a more abundant supply of fuel to both the above cities, from the immense beds of coal which lie dormant in many places of this county, and to carry lime from those parts in the track of the navigation where it abounded, to others where it was wanted. These carriages alone, it was believed, would have occasioned a considerable business on the canal;—industry, in different shapes, would have been aroused along its banks—lateral branches would have been brought in to join it—new products would have been raised through all the adjoining country—and easy intercourse would gradually have advanced the industry and fertility of extensive districts, at present wild and desolate, in a very high degree. On these considerations, so soon as the design of carrying this undertaking into execution was publicly advertised, there was an appearance of subscribers sufficient to advance the money necessary for the purpose. Surveys were appointed, and made along different tracks, by Mr. AINSLIE, as a surface surveyor, and by Messrs. GRIEVE and TAYLOR, as mineralogists; and very favourable reports were returned. But the war having commenced*, the attention of the public, and the capitals of monied men were diverted into a course diametrically opposite; and the hopes of this great national improvement being effectuated, are now more remote.

---

* War begun in 1793.

SECT. III.——WEEKLY MARKETS AND FAIRS.

IT is less the custom in this, than in some of the neighbouring counties, for husbandmen to carry their produce
immediately to market. A set of intermediate people
generally come between the grower and consumer. Fowls,
eggs, &c. are collected by itinerant dealers; and cheese,
and butter also, is either sold to such people, or to huck-
sters residing in towns. Grain of all kinds, likewise, is
sold either to dealers or to the consumer, by sample.
Hence there is little occasion for a concourse of the
country people with their commodities to market; and
the weekly markets in the different towns of the county
are much less attended than formerly. But there are
stated days in the week for market days in all the towns,
and the commodities of the country are still carried occasionally to these, and particularly to Glasgow, being the
chief consumpt, to which there is still a great resort for
the sale of consumable commodities, on the Wednesdays.
A weekly market, for the sale of horses and milch cows,
is also held there every Tuesday and Wednesday, for two
months, after the middle of January, yearly.

Though fairs are not so crowded as they are said to
have been in former times, there is still a great resort to
many of them, held at the towns and villages of the
county. Fairs are held at Carnwath, Carstairs, Lanark,
Carluke, Douglas, Lesmahagow, Stonehouse, Strathaven,
Hamilton, Shotts, Airdrie, Rutherglen, Kilbride, and
Glasgow. The times at which these fairs are held, may
be seen in all the almanacks. In them are sold, wool,
flax, lambs, cattle, horses, &c. At such meetings farm
servants are generally hired. The fairs are most frequent

in

in the ancient burghs of Rutherglen and Lanark. Near the village of Kilbride, three or four weekly markets, for the sale of sheep, are held on the Saturdays, the latter part of May and first part of June, yearly.

The customs exacted at weekly markets and fairs, are certain taxes which the magistrates of burghs and the proprietors of ancient baronies are, by the laws and customs of the nation, authorized to levy from the inhabitants of the country, who resort to such meetings to sell their commodities. This is one of the stigmas which the barbarous policy of the feudal system has affixed on the cultivators of land. These dues are always exacted insolently, and frequently with unjust rigour, by the meanest tax-gatherers; and are commonly paid by country people with indignant reluctance.

SECT. IV.——COMMERCE AND MANUFACTURES.

BEFORE the commencement of the eighteenth century, this county was but little known, either for its commerce or manufactures *. It was from the ports on the east coast that the intercourse of the kingdom with foreign countries was carried on; and this county had not a great quantity of commodities to exchange. It does not appear to have been remarkable, either for the quality of its wool, or for skill in manufacturing it, at any period of which tradition hands us down accounts. The manu-

---

* About the year 1668, WALTER GIBSON, an enterprising merchant of Glasgow, we are told, exported herrings to France in a Dutch bottom, and soon after was proprietor of three large ships.—*Denholm's History of Glasgow.*

facture

facture to which the inhabitants chiefly betook themselves, besides fabricating the necessary articles for domestic use, was that of linen. Before the Union, a considerable traffic was carried on, in all the towns of the county, in collecting linen yarn, and sending it to England, besides what was wrought up into cloth; and, fifty years ago, the women were become famous for making fine linen yarn. Various branches of the linen manufacture have been established, from time to time, in different parts of the county; many of the inhabitants have been instructed in the art of weaving; and that art has been admitted to be brought to a considerable degree of perfection. But the independent manufacturing establishments, in the central parts of the county, have all sunk, one after another; and, as the city of Glasgow has advanced in commerce and wealth, almost all the manufactures of the county have ultimately centred in it; and it being now the great mart of every commodity produced in the county, the commerce and manufactures of Glasgow comprehend the whole.

Before the Reformation, Glasgow seems to have been little concerned in any branch of secular industry *. Being the see of a rich archbishopric, its inhabitants were either ecclesiastics, or such as drew their support from the plentiful revenue raised by clerical address. But the hierarchy cherished in its bosom a viper, which, as soon as it gathered strength, was to sting its patron to death. The doctrines of the Reformation were greedily received in the western parts of Scotland; and many of the inhabitants of Glasgow became strongly tinctured with them. When the interest of the great concurred

---

* In the year 1718, the first ship belonging to Glasgow crossed the Atlantic.—*Denholm's History of Glasgow.*

with

with the religious zeal of the vulgar, the established religion was the easy victim of so powerful a combination: and the bold severe people, who had inveighed so loudly against the sloth, the knavery, the luxury, and the licentiousness of the clergy, resolute to act an opposite part, in every respect, assumed a new style of manners, and earnestly set about working themselves a livelihood, by what they deemed more creditable means. Accustomed to accommodate the occasional wants of the numbers who resorted to the metropolitan see, on a religious account, the inhabitants of Glasgow were somewhat inured to the habits of traffic and manufacture, which they now improved. The manufactures existing in the county were extended, some new ones introduced, and the internal commerce of the country gradually enlarged. Foreign commodities, to supply the demand of the neighbouring country, were imported; and fish, caught in the river or its estuary, and salted, with some other small articles, exported in change. The small capitals thus employed in the hands of people whose austere manners forbade the use of every amusement, and led to an unremitting attention to business, and the most rigid economy in the exercise of it, could not fail to increase. The spirit of Glasgow communicated itself, in some measure, all around, and the county was slowly advancing in wealth and industry, when the treaty of Union between the two kingdoms was concluded, and laid open a direct trade to all the British Colonies.

Those industrious merchants quickly availed themselves of this event: commerce made rapid progress*, and not

* In the year 1735, the number of vessels belonging to the Clyde, trading to foreign countries, was 67, and their tonnage 5600 tons.—Denholm's History of Glasgow.

only

only continued to enlarge the scale, on which all the former manufactures were conducted, but, from time to
time, introduced various others, which, from small be-
ginnings, became extensive and flourishing. In this
manner the commerce and manufactures of Glasgow advanced, and had arrived at a great degree of prosperity
at the commencement of the American war. This put a
stop to the tobacco trade, which had hitherto been the
principal source of wealth, and diverted the industry of
the country into a different channel. The war, which
for some years checked the progress of trade, was no
sooner ended, than the capitals, acquired by a long
course of successful industry, were sent in quest of new
employment. That wonderful exertion of human genius,
the machinery for spinning cotton, had now been invented, and brought to such a height, as to prepare the
material for the loom in much greater perfection, and
at a much cheaper rate, than heretofore. The manufactures of Lancashire, which, before, were much employed on cotton, were extended and improved, by means
of yarn obtained from the new machines; and, as they
advanced in refinement, had begun to vie with the elegant productions of the eastern looms. This inflamed
the emulation of Glasgow. All the new machinery necessary for the cotton manufacture were introduced into
the county; large quantities of the raw material were
imported; and not only all the different kinds of goods,
formerly made of linen, but imitations of the various
manufactures of India, formed from that material. The
art of dyeing was much improved; and a durable tinc-
ture, of various hues, given to cotton. The printing of
cloth, too, made great advancement; and large quantities of cotton garments, of elegant patterns, were executed and exported, with other goods, to the different
markets

markets of Europe and America. But the conduct of
this manufacture was not confined to people of capital
and established credit. The banks were become numerous ; and, if they were not all possessed of large capitals, from the manner in which they had, for a considerable time, been conducted, they had at least obtained
very extensive credit ; which, from the prospect of
greater emolument, they liberally parcelled out among
the manufacturers, and strove with one another who
should have the greatest share in the business. By such
means, adventurers, without stock or experience, were
enabled to contend with those who had both ; and the
contention of so many individuals, to enlarge their own
particular concerns, raised the wages of every branch of
manufacture to an immoderate height. The numbers of
people already engaged in the manufacturing occupations
being insufficient, new ones flocked to the different
works, from all quarters and all employments ; the males
to the loom, the females to the flowering of muslins, &c.
Even those who, from their tender years and weak capa-
cities, were hitherto reckoned unfit for any kind of profitable labour, found employment in the cotton mills,
and other small works; so that there were perhaps few
families in the county, some part of which was not engaged
in this extensive manufacture. Of 126,000 people,
which was about the population in the year 1792, if we
comprehend all classes, from the master manufacturer to
the child begun to plant the teeth of cotton cards, the
number thus employed must have been 60,000, or upwards. When the high wages are considered, the annual value of all these people's labour must have amounted
to a large sum : but it would be scarcely possible to make
a fair computation of it, the manufacturers of Glasgow
not being confined to the county, but having extended
over

over a considerable part of Scotland. It has been said, that the operation of these manufactures on the materials imported in a year, adued, at least, a million to their value; and when the profits of the manufacturer, residing in Glasgow, are taken into view, it might have been supposed that so much should have centred in the county, as would have tended greatly to enrich it, and better the condition of all classes of the inhabitants. But a long course of prosperity having banished the frugal habits by which former wealth was amassed, an opposite style of manners had taken place, and pervaded all ranks; and upon the approach of the war, this phantom of prosperity seemed to have vanished, and, " like the " baseless fabric of a vision, left not one trace be- " hind."

Commerce suffered a severe shock, goods were accumulated in the hands of the manufacturers, frequent bankruptcies occurred, and great numbers of the operative people were thrown idle, and without bread. The consternation was general and great, but not of long duration. The surplus artificers betook themselves to the army, or emigrated; and the business of manufacture, being disencumbered of most of the rash unexperienced adventurers, was left to the conduct of those whose capitals, abilities, and professional knowledge, were more adequate to the employment. Those expert people did not remain inactive; but having discovered new vents for their goods, began to act on surer grounds than before; and the manufacture has been steadily carried on with great success ever since.

The cotton manufacture has gone on since last publication, though with various success, and ever fluctuating, but still been upon the increase. The quantity of cotton wool imported into the Clyde in the year 1804, was

CLYDESDALE.]   o   39,000

39,000 bags and pockets, which averaging 200*lb*. each, will be 7,800,000. By the opinion of the best informed, this cotton, spun, will be worth 4*s*. and 6*d*. per *lb*. at an average. This makes the value in yarn 1,755,000*l*. sterling. And supposing the value to be doubled by the operations of weaving, bleaching, printing, &c. the value of the cotton manufactures in the west of Scotland, will be 3,510,000*l*. annually. This, however, is a vague computation for different reasons, particularly, as much cotton wool is always re-exported from the Clyde to Lancashire, and much yarn returned from thence to Glasgow, the quantity of neither of which can be easily ascertained.

Though the magnitude of the cotton manufacture, in a general view, is such, that it,

" Like Aaron's serpent swallows up the rest,"

there is a great variety of others in the county, and more particularly at and around Glasgow, in which large capitals and great numbers of labouring hands are employed.

The manufacture of cast iron goods, at the iron works, has been mentioned. By these are made, cannon, balls, mortars, shells, grates, stoves, pots, and a long list of different utensils. Smith-work, in malleable iron, is also done to a considerable extent; and buildings are now erecting on the banks of North Calder, near Airdrie, for mills to split iron, where it is proposed to manufacture different kinds of Birmingham goods.

The tanning of leather, and the manufacture of boots, shoes, and saddlery, is carried on to a considerable extent, in different parts of the county.

The linen manufacture is still carried on, though on a more contracted scale; and great quantities of nuns thread

thread are manufactured. The inkle manufacture was early introduced into Glasgow, and is now pretty extensive. Carpets, and other coarse woollen goods, and hats for domestic and foreign sale, are also manufactured.

Not only manufactures of bricks, tiles, and the coarser kinds of pottery, but of delft and stone ware, bottle and flint glass, for all the different purposes, have been long exercised.

To this list may be added the manufactures of ropes, lines, and cordage of all kinds; of soap, candles, sugar-boiling; the founding of printers' types, printing, dyeing, bleaching, printing of garments, &c. &c.

To give a more minute detail of all the different branches of manufacture exercised in the county, would be as difficult as needless. But a just statement of their extent, of the capital they occupy, the profits which accrue, and the numbers which they employ and support, might be more important, if it could be given. Any attempt of that kind, however, would be merely conjectural, and might tend more to mislead than inform. The reader must therefore be left to form his own conceptions on the subject.

It ought here to be observed, that a number of the inhabitants of Glasgow have been remarkably zealous in making science an auxiliary to the arts. Chemistry, and other branches of experimental philosophy, have been studied with great assiduity, and applied with success to the advancement of different manufactures; and in every department much professional knowledge, much spirited enterprize and persevering industry, has always been displayed. Unhappily the demon of speculation spreads his fascinating allurements, and too often catches the unguarded. Then the moral and intellectual feelings are benumbed, and the speculator is alive only in pursuit of

the

the favourite object. If he succeeds, he enriches himself at the expense of the public, and tempts others to follow his example. The unsuccessful not only ruin themselves, but all who place confidence in them. Little more than a year ago, many bankruptcies, the consequences of such projects, occurred, to an astonishing amount.

The ports on the Clyde, through which the foreign commerce of Glasgow is transacted, lie in another county, in the Report of which the tonnage of the shipping employed will probably be given, by which the reader will judge of its extent*.

This great extent of commerce and manufacture, while it has, on the one hand, tended to promote agriculture, by enlarging the demand for its various productions, has, on the other, proved a great check to the gradual progress of agricultural improvement, and prevented that increase of land produce which a growing population required. It has withdrawn the capital, and allured the most enterprizing inhabitants from the cultivation of land to the pursuit of more splendid projects; it has seduced the peasantry from their residence in the country and the labour of the fields, to seek a life of greater ease and enjoyment in towns and manufacturing villages. The

---

* Besides the trade of Glasgow, Greenock, situated 24 miles farther down the Clyde, has quickly risen from a group of fishers' huts to a large commercial place. The number of vessels trading from the Clyde, belonging to both towns, was in

1783	386	Tonnage	22,896
1790	476	Tonnage	46,581

And besides the vessels belonging to the Clyde, those of other nations, especially the American States, come to that port, and exchange their commodities for the manufactures of Glasgow. The number of vessels trading to and from the Clyde in 1796, were 3095; their burthen 238,790 tons.—*Denholm's Hist.*

new

new modes of life introduced into the clusters of artificers, among whom the profligate and the dissipated generally make a part, have spread their contagion among
the lower orders of the people, tending not only to ener-
vate the body, and disqualify it for laborious exertions,
but to contaminate the morals, and destroy that simplicity and decency of manners which is their most important quality, either with respect to their own interest or
that of society. The tempting encouragement of manufacture has thus diminished the number, and raised the
wages of labourers to a great pitch. While the advance
of rent, of wages, &c. has inflamed the expense of cultivating corn, the facility of commerce, and the encouragement of bounties, enabled the merchant to import it
at a lower rate than it could be raised in a cold and barren country. This, accompanied with the great rise on
all kinds of provisions obtained from pasture land, induced many occupiers of land to betake to grazing, which
occasioned much land to be left in grass before it had undergone such culture as was necessary to make it produce
grass; by which the present produce of the county was
diminished, and its future improvement retarded; for
when land, naturally fertile, or made so by cultivation,
is laid in grass, it is soon covered with a close turf, and
the quality of the herbage improves while it is continued
in pasture; and when it is again brought into tillage, it
repays the patience of the community with an increased
produce of grain: but on poor lands, and especially such
as have a wet bottom, the herbage becomes coarser, and
less in quantity, the longer it lies in pasture; and the
soil becoming wilder, the difficulty of improving it is increased. The truth of this assertion, and of the inference implied, will be acknowledged by the most superficial observer, who has had occasion to mark the state of

o 3   the

the poorer soils of this country for the last 30 years;
where numerous instances occur, of fields, the surface of
which were almost naked about the beginning of the
above period, that now, by the ordinary exertions of
common farmers, with no manure but the farm-dung,
and a little lime carried annually, are bearing tolerably
good pasture; whereas such similar fields as have been
neglected, bear little esculent herbage, and, at the present price of all the means of cultivation, could not be
now improved without loss.

SECT. V.——THE POOR.

ANY attempt to give a statement of the number of
those who derive their sustenance from the charity of
others, or of the amount of what is thus bestowed, would
be very defective, and give no just idea of the subject.
As to the first, the numbers of the poor are certainly
very great, and seem to have increased, as the extension
of manufacture has afforded more ample support to the
industrious. After what has been said in the former
parts of this Report, it will be needless to offer a solution
of this seeming paradox. Besides all those that are to
be found on the different parish lists, or are entertained
in charity houses, &c. numbers of mendicants swarm from
the populous towns of this and the neighbouring county,
over all the country, and extort charity by all the arts
known to people of that profession. Many of the mo-
dest, who have sunk under the pressure of misfortune,
are supported by private charity. With respect to the
support of the poor, besides the assessments, and other ordinary funds of the different parishes, and the charitable
institutions, of which there are many in the county (the
most

most considerable are in Glasgow, and an account of them may be seen in Sir JOHN SINCLAIR's Statistical Account of Scotland, vol. v. p. 518), very liberal contributions are made in all calamitous emergencies, whether general or particular. The alms extorted by the practice of mendi-city must surely be considerable; and the sweeter obla-tions of the feeling heart, which impels the right hand to do what the left hand knoweth not, far from trifling. But great as they may be, the poor ought certainly to be supported, by whatever means their poverty has been brought about. Happy would it be, if some expedient could be devised to relieve the wants and soften the dis-tress of the children of misfortune, in a frugal way, without debasing the minds of those who were supported to the abject state of beggary, or inducing idleness among others, on the prospect of such support!

The numerous Friendly Societies now instituted in this county, for the relief of their own members in distress, so far as they go, bid the fairest for the attainment of this purpose. The wisdom of the regulations made for ma-naging the affairs of these societies, and the integrity, frugality, and good effect, with which they are adminis-tered, is such, that it would do no discredit to any of the higher classes, to whom more important trusts have been committed, to have their conduct compared to that of the managers of the Friendly Societies of Clydesdale. Every member of any of these societies contributes a small pit-tance monthly or quarterly to the general stock, and re-ceives from it, in all cases of real distress, a comfortable support. It seems to be no less the interest than the duty of people of the higher orders, to lend their aid to make this mode of supporting the poor general; since it would perhaps contribute to the preservation of their morals, as

well

well as to their comfort in distress. People might not only give donations, according to their ability, to one of these societies, but they might oblige the servants whom they admitted into their families, to become members of it. The more general it were to enter into such societies, it would become the more discreditable to neglect it ; and perhaps there would fewer remain with that baseness of disposition, which prompts people, either from selfish or wanton motives, to injure the property of others; for, it is believed, there are few or no instances of the regular adherents of these societies being caught in crimes.

SECT. VI.——POPULATION.

WHEN this was last published, a table of the population, extracted chiefly from Sir JOHN SINCLAIR's Statistical Account, and the deficient parishes, made up by other means, was inserted ; but the people having been enumerated by authority of an Act of Parliament, in May 1802, we are thus furnished with a later list, which is stated in the following table, and contrasted with the former.

TABLE

TABLE of POPULATION, from the Returns to Parliament under Act 41st of GEO. III. compared with the former.

Parishes.	Statist. Account. Souls.	Late Enumeration.		
		Males.	Females.	Total.
UPPER WARD.				
Biggar, - - - -	937	555	661	1216
Carluke, - - - -	1730	866	890	1756
Carmichael, - - -	781	403	429	832
Carnwath, - - -	3000	1297	1383	2680
Carstairs, - - -	1200	407	492	899
Covington, - - -	484	216	240	456
Crawford, - - -	1490	848	823	1671
Crawfordjohn, - -	590	337	375	712
Coulter, - - -	326	182	187	369
Dolphington, - -	207	117	114	231
Douglas, - - -	1715	759	971	1730
Dunsyre, - - -	360	176	176	352
Lamington, - - -	417	189	186	375
Lanark, - - -	4751	2180	2512	4692
Lesmahagow, - -	2996	1560	1510	3070
Libberton, - - -	730	322	384	706
Pettinain, - - -	386	200	230	430
Simonton, - - -	264	133	175	308
Walston, - - -	427	165	218	383
Wiston and Roberton, -	740	368	389	757
MIDDLE WARD.				
Avondale, - - -	3343	1722	1901	3623
Blantire, - - -	1040	785	966	1751
Bothwel, - - -	2707	1470	1547	3017
Cambuslang, - -	1288	787	771	1558
Cambusnethan, - -	1684	888	1084	1972
Dalserf (no return made),	1100	550	550	1100
Dalziel, - - -	470	305	306	611
Glasford, - - -	788	466	487	953
Hamilton, - - -	5017	2686	3222	5908
Kilbryde, - - -	2359	1119	1211	2330
Monkland East, - -	3566	2184	2429	4613
Monkland West, - -	4000	2006	2000	4006
Shotts, - - -	2041	1007	1120	2127
Stonehouse, - -	960	584	675	1259
LOWER WARD.				
Cadder, - - -	1767	992	1128	2120
Carmunnock, - -	600	332	368	700
Cathcart, part of (conjectured),	130	65	65	130
Glasgow city and suburbs in first list, alone the second,	61945	20446	26333	46,779
Barony of ditto, besides suburbs in first list, including suburbs the second,	3093	12717	13993	26,710
Gorbals, included with Glasgow in first list,	...	1844	2052	3896
Govan, - - -	2518	3277	3424	6701
Rutherglen, - -	1860	1200	1237	2437
	125,827	68,712	79,214	147,926

CHAP.

CHAP. XII.

OBSTACLES WHICH IMPEDE IMPROVEMENTS IN AGRICULTURE,

INCLUDING GENERAL OBSERVATIONS ON AGRICULTURAL LEGISLATION AND POLICE.

HERE a crowd of ideas rush on the mind, and make it doubtful whether it be proper to stop or to proceed. The legislature of a great nation, which has too long disregarded the culture of its soil, its most important interest, or mistaken the means of promoting it, has at length been convinced of the necessity of paying more attention to it, and has chosen, from the most distinguished classes of its citizens, a Board, to preside over the National Agriculture, and examine by what means its success might best be advanced. This respectable Board, in order to call the attention of the public more fully to a subject in which people of all ranks and conditions are interested, has very properly solicited information from every corner. In such a case, it would be affronting the Board of Agriculture, and the public, who expect ample information through this channel, to amuse them with a few incidental inconveniencies attending husbandry in a particular district. Such local inconveniencies have no considerable effect on the state of the national agriculture, and would be easily obviated by experience and industry, so far as circumstances admitted, if the more general obstacles were removed.

But if it is useless and unsatisfactory to treat of these frivolous topics, alas! it is much to be feared, it would be equally vain to attempt to state those more important and
formidable

formidable obstacles, which were first founded by ignorance and barbarity, and have since been reared and confirmed by prejudice. From the earliest periods of social
civilization, contemplative men, of all ages and nations,
have discovered the immense importance of agriculture,
and exerted the powers of reason and eloquence to re-
commend the practice of it to their countrymen. Though
these have not been wanting in this island, though the
safety of property, under the regular administration of
law, and the high demand for every kind of land produce, have long afforded encouragement to agriculture,
far superior to what exists among the surrounding nations, yet such have been the ambitious views of statesmen, with which the public have always too eagerly concurred, and sometimes spurred on, that the pursuit of
war and conquest, colonization and commerce, has been
preferred to the fundamental support of the nation—the
cultivation of the soil; and the Reports returned to the
Board in 1793 and 1794, bear ample witness, that agriculture is left far behind by other industrious arts. Since
such has been the case, in spite of several intervals of
tranquillity, in which the nation has had leisure to attend to its internal resources, and strong lessons to make
them its peculiar care, what hope remains that the feeble
representations of an obscure individual, or even of any
number who may happen to concur, can now have any
effect, at a time when we are informed by the best authority, that we have to struggle for our existence as a people!

> O mortals, mortals! when will you, content
> With Nature's bounty, that, in fuller flow,
> Still as your labours open up its sources,
> Abundant gushes o'er the happy world;
> When will you banish violence and outrage,
> To dwell with beasts of prey in woods and deserts?

THOMS. CORIOL.

But,

But, however hopeless the execution of this task may be, since it has been undertaken, it has become a duty; and therefore the great obstacles to improvement in agriculture shall here be shortly stated: and, though the Writer does not mean to confine himself to such as are peculiar to this county, which, strictly speaking, would amount to nothing, he will dwell chiefly on those that are common to this and other surrounding provinces in similar circumstances. But, before we proceed, it will be necessary to premise, that there are circumstances connected with the situation of Britain, which tend to magnify existing obstacles; or, in other words, these circumstances demand that agriculture should have more support and encouragement in Britain than may be necessary in many other countries. The circumstances alluded to are, 1*st*, The climate and soil; 2*d*, The inadequacy of the produce of agriculture, in its present state, to supply the wants of the inhabitants.

1*st*, The climate of Great Britain, from its insular situation, is cold and unsteady. Surrounded by a vast ocean, its vapours frequently hang over the country, and exclude the maturing influence of the sun. This is more particularly the case with the northern half of the island, where the inconstancy and inclemency of the weather frequently not only interrupt the labours of the husbandman, but also disappoint his harvest expectations, in spite of all his diligence. Under such a sky, it is not to be expected that the soil can be more favourable. Accordingly, we find its fertility is kept up only by dint of industry and attention *.

2*dly*,

---

* The frequent repetitions of rain and storm condense the soil; and the water lodging in it unites its particles into a cohesion, impermeable to the roots of plants. Much of the best soil too, is washed away in torrents.

2dly, The land produce of Britain seems to be short of the demand of its own inhabitants. The Writer recollects, that, several years ago, an inquiry was made, by order of the Privy Council, into the state of imports and exports of corn, to and from all the ports of Britain: The result was, that, for an average of 18 years immediately preceding the time of the inquiry, the imports exceeded the exports considerably; whereas, for an average of 18 years previous to that period, the balance stood greatly on the opposite side. In order to have stated this more accurately than from loose recollection, application was made at the custom-house of Edinburgh for the sight of a copy of the Report made to the Privy Council, which was promised, but has not yet been obtained*. Should it be got, an extract shall be given in an Appendix. But, if the corn raised in Britain was not sufficient to feed all its inhabitants more than 20 years ago, there is reason to suspect, that the deficiency has been much greater of late, when such vast imports from all quarters of the world have been made.

But, whatever may be the case in general, the deficiency, both of animal and vegetable food, for the support of the inhabitants, produced in this county, is very considerable. A few fatted calves, some sheep and lambs, butter and cheese, go from the upper parts of the county to the Edinburgh market; but the animals for slaughter, from the more western and northern parts of Scotland, and the salted beef, pork, and butter, from Ireland, and cheese from different counties of England, con-

---

torrents. Hence the difficulty of restoring it by cultivation is greater than in a mild climate: and the state of the weather is frequently unfavourable for that purpose.

* This was never got, but is not now necessary.

sumed

sumed in Glasgow and its neighbourhood, are much more considerable. The neighbourhood of Glasgow being the best corn market in the country, there is scarcely any of the corn raised in the county carried out of it; and much from other quarters is annually consumed in it: but there is no rule by which the quantity of imported corn can be accurately ascertained. In a late corn law, which prescribes regulations for the importation and exportation, the county of Lanark is joined with those of Renfrew, Dumbarton, Bute, and Argyle, each of which, and especially the first, requires always a great deal of foreign supply. But, let us suppose that the consumption of Glasgow is equal to one-third of the whole import. The oats and oat-meal imported into the Clyde, in the year 1790, was 107,000 bolls, of eight stones Dutch each boll. The same year, there were brought along the canal to Glasgow, &c. 164,000 bolls of oats, oat-meal, wheat, barley, and pease. But a boll of wheat, barley, or pease, will produce much more than eight stones of meal; let us therefore add one-fourth more to this, or 41,000 bolls, which makes the total import 312,000, and the consumption of Glasgow and its neighbourhood, according to the above supposition, 104,000 bolls, or 832,000 stones Dutch. The importation has, some years since, been still more considerable; but this increase may, in some measure, be imputed to the number of cavalry quartered in the country.

It is equally difficult to compute what the whole consumption of the county may be annually. What is consumed by the distillery, by brewing, and by the great number of horses kept for all different purposes, cannot be ascertained with any degree of accuracy that could be depended on. It is presumed, that $2\frac{1}{2}$ bolls, or 20 stones of all kinds of meal and flour, with the quantity of potatoes

tatoes commonly used (the most part of which are raised in the county), may serve each person, at an average, for farinaceous food, through the year; and, supposing the population, as above stated, about 126,000, the total consumption (by the human race) will be 315,000 bolls, or 2,520,000 stones, nearly one-third of which is imported from different parts.

This deficiency of food, which seems gradually to be increasing, it would appear, is not peculiar to the large province with which this county is connected, but, if what is above stated be agreeable to fact, is, in some degree, the general fate of Britain, and, therefore, merits the most serious consideration. Whatever the skill and activity of a nation may be in commerce, or in the refinements of manufacture, it is surely alarming to be generally in want of food; as it must put the independence and continued prosperity of such a nation on a precarious footing. Food is the principal and natural wages of labour, the great incitement to industry, and the cause of the increase of an industrious population [*]. A sufficiency of food, therefore, must be the strength of a state; and the means used for producing it the most important employment. In some of the transactions of the American Congress, published in the newspapers a

---

[*] To avoid lengthening out this Report with numerous quotations, the reader is here referred to a work of great merit, well known to the public, viz. An Inquiry into the Nature and Causes of the Wealth of Nations, by the late Dr. SMITH, Book I. Chap. XI. Part 1st and 2d. This very intelligent and respectable author, though he may, perhaps, have erred in some particulars, which, in the execution of a task so arduous, is not surprising, has favoured the public with a greater number of just and liberal principles of political economy than any other book in the language contains. It is hoped, therefore, it will be allowable to resort to such authority; and it may again be necessary to do so in the sequel.

few

few years ago, here quoted from memory, the Ameri-
cans, comparing themselves with the manufacturing nations of Europe, say, " We are the masters and employ-
" ers of manufacturing people ; they are our labouring
" servants. In our superabundance of provisions, &c.
" we possess the wages of their labour ; we can retain
" these ; we can retrench our superfluities, and abstain
" from employing them ; but they cannot live without
" our employment."

Having thus premised the strong necessity of removing
every obstacle which stands in the way of improvement in
agriculture, and of using every possible means to promote
its advancement, we come now to state these obstacles ;
and, it is hoped, what has been advanced will justify
what follows.

The first obstacle which shall here be mentioned, is
the little respect which has been shown to husbandry, and
the marks of degradation and servitude which the laws
and customs of the country have attached to the profession. A number of instances, sufficient to illustrate this,
have been given in the preceding parts of this Report ;
and to these the reader shall be referred, without trou-
bling him with a recapitulation. From an employment
of such importance, and from which the necessities of
the nation demand so many additional exertions, every
thing that tends to damp the spirits, or fetter the activity,
should be removed. Without taking time to collect authorities, or multiply arguments to show how unfavourable servitude, or whatsoever borders upon it, is to the
success of agriculture, we shall refer to the sentiments of
the respectable author just now mentioned, from whom
we take the following short quotation : " In ancient Italy,
" how much the cultivation of corn degenerated, when
" it fell under the management of slaves, is remarked
                                            " by

" by both PLINY and COLUMELLA*," &c. &c. If depriving
the cultivators of land of their freedom, tends to depress
cultivation, then it must surely follow, that, to put them
upon the most liberal and respectable footing, would
greatly forward the success of that art.

The unhappy jealousy which subsists between landed
and manufacturing people, is another obstacle to the success of agriculture. The Author of this, in a former
work †, has endeavoured to show how ill-founded this
jealousy is, and how strictly the interest of both classes
is united. Our present business is to explain the bad effects of the want of this union. Landed people have
beheld, with some degree of envy, numbers rising into
consequence by the effects of industry, and have wished
to lay that industry under contribution, by restraining
laws, intended to raise the price of provision. Manufacturing people, on the other hand, have been making
perpetual struggles to counteract this; and, by loud cla-
mours of the danger of the poor being starved, have obtained certain relaxations of those laws. Hence has
arisen that contradictory jumble of statutes and regulations, known by the name of the corn laws, and the offi-
cious interference of the executive government, in all
pretended emergencies. The part which land-holders
have taken in forming these laws, was, probably, with a
view to increase their own revenue, rather than to improve the condition of husbandmen, or promote agriculture. It would seem, however, to have been without
effect. Rents, indeed, have greatly risen; but the restraining laws have not been the cause. The price of

---

* SMITH's Wealth of Nations, Book III. Chap. II.

† NAISMITH's Thoughts on Industry. Edinburgh, printed 1790, Book
III. passim.

CLYDESDALE.]   P   butchers'

butchers' meat, and butter, never rose to a great pitch,
till after liberty was given to import live cattle, salted
beef, and butter, duty free, from Ireland; and the price
of corn is not advanced. It is true, indeed, as appears
by the table of the fiar prices of grain, given in page 173,
there is a small advance of the average money price of
each 15 years above the former; but, in the last period,
there were two years of dearth. Besides, though there
is a small rise in the money price of corn, its real price
is greatly fallen. That the value of money is greatly
fallen, will not be denied. Whether this has been occasioned by the increased quantity of circulating paper, or
by what other means, we stop not to inquire. The judicious author lately quoted, observes, that labour and
corn are the best measures of the value of one another.
Let us try the price of corn by this rule. About the
year 1760, and for some years after, the wages of a labouring man were 8d. a day. The average price of meal,
in that period, is 10d. per peck; the wages of a day's
labour, therefore, would not purchase a peck of oat-
meal. The average price of a peck of meal, in the last
period, is about 12½d., and the wages of a day's labour
being 1s. 6d. will nearly purchase a peck and a half.
Hence the value of corn is reduced more than one-third
since the year 1760. The real price of beef and butter,
the importation of which from Ireland, a plentiful country, is unrestrained, is, however, advanced. In the year
1760, the wages of a day's labour would have purchased
at least three pounds of beef, or two pounds of butter;
at present it will not purchase much more than two
pounds of beef, and not one pound and a half of butter.
The real price of the provisions obtained from pasture is,
therefore, advanced nearly as much as that of those obtained from tillage is depressed.

From

From this view of the matter, it is obvious, that all the attempts of landed gentlemen to obtain partial laws, for their own emolument, have been without effect. The interest of husbandmen, and the general cause of agriculture, however, have suffered in the struggle. If the commerce in provisions had been at all times free, without restraints or bounties, importers would have been cautious to import no more than they had a prospect of selling with profit; and the provisions produced at home would always have had as much advantage in the market, over the foreign provisions, as the expense the latter cost in importing, &c. and the damage they often suffer by sea carriage. Hence, in times of scarcity, the husbandmen of the country would have got a higher price for such provisions as they had to sell, which, though perhaps, not adequate to the deficiency of quantity, would have, in part, compensated for it. But, according to the present economy, this is not the case. The crops of 1794 and 1795 were much more productive, through a great part of Scotland, than that of 1796, yet the prices of corn were continually rising during the two first productive years, and fell very low before the crop of the last less productive year was consumed, the very best oats having sold, in the beginning of this summer, at 12s. per boll, and under. The present summer, 1797, has been the most cold and barren one ever remembered. The crop has been now reaped, and found defective; yet this deficiency has hitherto had very little effect on the market price of corn. Husbandmen have nothing to say in the matter; and, besides, they are, as has been already shown, unequal to such struggles. Landed gentlemen, the champions of the cause on the one side, are not equal to their opponents. Besides, the latter have the popular clamour on their side, and, apparently, the

cause of humanity. A minister, therefore, finds it necessary to lean to this side; and, in order to gain popu-
larity, if simply opening the ports does not satisfy, he
grants large bounties to encourage importation.

Thus, it appears, that, if any party has gained by this
vain contest, it has been the manufacturing interest, or
the consumers. But it would have been no worse for
them, if the cause of contest had never existed; if importation and exportation of corn had been always free,
unassisted and unrestrained. This was the highest of
their demand; and it was all they could reasonably demand. Had this been the case, so soon as an increase of
industrious people, and the more expensive manner of
living, induced by the increase of wealth, had raised the
demand for provisions above what the present state of
culture could supply, the husbandmen of the country,
having an advantage over those of other countries, equal
to the whole expense of importation, would have been
excited to make new improvements, to answer the increasing demand; and would have, probably, succeeded.
Consumers would thus have had the sure pledge of a
more certain and regular supply than can be expected
from the operations of foreign commerce.

The Writer would have been ashamed to have dwelt
thus long on a topic so obvious, if it had not been that,
obvious as it is, the principle does not seem to be generally admitted. He therefore thinks it his duty to take
this opportunity of bringing it once more under the re-
view of the public; and besides, it seemed necessary to
say so much, in order to illustrate the position, that the
discordance between the landed and manufacturing interests had been an obstacle to improvement in agriculture. He is of opinion, that this discordance has the
same effect in other instances; but he despairs of being
able

able to explain them in such a manner as to be fully comprehended. He must therefore be content with observing, that of all the people whom the allurements of manufacture have abstracted from the labours of the field, very few ever now return to lend occasional aid. The connexion is entirely broken off, except when they become outcasts from manufacture, and issue forth to beg their bread among the farm-houses of the country.

The over-stretched rent of land, the scarcity of labourers, and the high price of their wages, taken together, make a formidable obstruction to improvement in agriculture. The first has been already noticed in this Report, pages 68 and 82, the last in pages 61 and 196. The rapid rise of land-rent seems rather, at first, to have originated in a kind of frenzy among farmers, than any premeditated design of the landlords; but the latter quickly availed themselves of that frenzy, and have certainly carried the matter too far, with regard to the general interest of agriculture. This rise of rent, it would appear, must at length come to a period. It has been shown (page 171), that the increase of rent has not been supported by an increase in the price of corn; and from what is stated (pages 204, 205), there is some reason to suspect it is not much more supported by the increase of quantity. The scarcity of labourers, and the high price of their wages, adds to the burden of high rent, which make the husbandman's prospect of profit small and precarious, prevents the increase of his stock, and thus checks his spirit, and weakens his energy in the cultivation of his farm.

The three obstacles above enumerated, are the cause of a fourth, not less considerable; that is, the deficiency of the stock employed in agriculture for the purpose of carrying the improvement of the country to the necessary

P 3   extent.

extent. The low estimation in which husbandmen are held, must, no doubt, sometimes prevent men of generous souls from embracing the profession. The advantages which commerce and manufactures have gained over agriculture, have led much capital and enterprize from the latter to the former. Much agricultural capital was consumed in the numerous bankruptcies which succeeded soon after the great rise of rents took place. At that time, many cautious husbandmen withdrew from the employment, and carried their capitals along with them; and there is little hope that, in the present state of things, any considerable quantity will return from other employments to the support of agriculture. Agricultural stock has, no doubt, been increasing again, of late years, in the hands of some of the most expert and successful husbandmen; but it has been losing its efficacy in a much greater degree. It is observed by husbandmen advanced in life, that in no case the efficacy of capital is so greatly sunk as in the stocking of a farm; it being generally believed that, from the difference of the expense of all kinds of utensils and implements, the difference in the value of labouring horses, and other live stock, &c. 20*l.* would have gone as far, 40 years ago, as 100*l.* will do now.

But the principal obstacle to carrying improvements in agriculture to the greatest height of which the country is capable, is the great extent of land held in property by great land-holders, the shortness of the leases given to tenants, and the frequent practice of shifting the tenants at the end of every lease. Loaded with all the hardships and difficulties, enumerated and described in different parts of this work, under which farmers labour, their exertions to improve their farms can seldom be very vigorous; and they are generally too solicitous to reap the
whole

whole advantage of such meliorations as they are able to make, which they sometimes carry so far, as not only to exhaust the land, but to hurt themselves. From the commencement of the lease they begin to look forward to the end of it, and hopeless of any preference to a new one, regulate their conduct accordingly. While the land is cultivated by the labour and capital of those who have no interest in its permanent improvement; while the landlord impatiently expects the fall of leases, that he may advance his rental; while the farmer makes it his study, as far as the articles of lease or the indolence of the landlord will permit, to disappoint this expectation, any great degree of increase in the fertility of the country can never happen. These jarring interests must be reconciled, and the influence of the landholder and the husbandman collected into one focus, before the face of the country can be effectually improved, and a superior quantity of food from our native soil obtained.

The engrossing of a great extent of land into single farms, is another obstacle to the general improvement of the country. The question of the comparative advantage between moderate and great farms, has been much canvassed; and many advocates have appeared in favour of great farms. Of late, however, the tide of public opinion seems to have turned a little to the side of moderate farms. But the Author of this having, in a work to which he has already referred, examined this subject at considerable length, he begs leave to refer the reader again to the perusal of it*, which he hopes will convince the candid mind, that a country will be more advantageously cultivated, when a sufficient number of those em-

---

* NAISMITH's Thoughts on Industry, page 609, and onwards.

ployed

ployed in the cultivation are engaged, by their own interest, to pursue their labours with diligence and attention, than it could be by hirelings.

But, say the advocates for extensive farms, the barnyards of great farmers are the best granaries for preserving a regular supply of provision. Were the country cultivated by small farmers, grain would be rushed in, and glut the market at one time, so as to produce a low price and immoderate consumption; and hence scarcity and dearth would ensue; whereas the great farmer brings his grain to market only when he finds there is a ready demand for it; and thus the provision of the country is better husbanded. It does not appear, however, that moderate farmers should necessarily be less prudent than great ones, and that all the small farmers of a district should bring their corn to market at the same time; nor do we find the market prices less fluctuating, as more of the country has fallen into the hands of great farmers. But the spirit of speculation which, in the mercantile world, has long been the scourge of society, has now spread its contagion among other classes. Great farmers become bold speculators, and, deaf to the wailings of the suffering poor, have in some instances pushed their speculative views too far, and become sufferers in a falling market. A greater number of moderate farmers, having it much less in their power to avail themselves by speculations, are less disposed to attempt speculating, and markets are left to their natural course. A periodical publication, in every body's hands (the Farmer's Magazine), sneers at moderate farms; but he wishes to command belief of the superiority of large ones by dint of authority, not to convince by fair argument. He does not recollect, that those lands on which large farmers succeed, have been fitted

to

to the purpose of the great farmer by the long continued application of a number of small farmers on the different parts of the large farm.

The only obstacle of inferior note, which shall here be noticed, is the effect of the game-laws. These are said to be less oppressive in Scotland than in England, it having been settled by some late decisions of the Court of Session, that game is property. But still it is the property of the proprietor of the land, not of the husbandman, on the produce of whose industry it is fed. The latter must not kill the hare that spoils his young orchard or his kitchen garden ; nay, he would be ill looked upon by all the fox-hunters of the neighbourhood, if he were to destroy the robber of his hen-roost, though caught in the act : but a pack of hounds, with a dozen of men on horseback after them, driving with the fury of Bacchanalians, may penetrate his enclosures, trample down his fences, poach over the ground softened with the winter's rain, destroying his young wheat and sown grass, and do more harm in an hour than all the hares and foxes would have done in a year. The fowler, too, though a less destructive animal, shows equal disregard to the property of the husbandman. Every one who thinks it genteel to be a sportsman, and can purchase a game license, sallies forth, wherever the permission or indifference of land proprietors give him sufferance, as soon as the time prescribed by the game laws arrives, for the destruction of grouse or partridges. In vain the timid birds use all the speed of foot and wing to elude the chase. The pointers, staunch to the cause of blood, and their eager followers, still pursue. Every thing then must give way to the ardour of the sport. The annoyance of the peaceful flocks on the heathy mountains, or the breaking down of fences, and trampling on the corns in cultivated lands, are nothing !

thing! However cautious and reserved sportsmen may be in the pursuit of their game, the husbandman always suffers less or more; but the gentlemen have had their sport, and the rustic must be silent.

It is observed, that birds of game have diminished in numbers, as the laws to preserve them for the sole enjoyment of legal sportsmen have been made stricter. This has been imputed to the resentment of husbandmen, who, it would seem, are allowed to have some reason to feel the inequality of those laws. But the decrease in the numbers of game, it is obvious, is owing to another cause. Game can only abound where the industry of man is not exerted. Animals in a wild state are harassed by the progress of improvement: they are obliged to shun the scenes of cultivation; their empire is narrowed; their opportunities of propagating, disturbed by numberless accidents; and thus their numbers are gradually diminished, without any intentional interference of husbandmen. On the contrary, the continued existence of birds of game in considerable numbers, is the strongest proof of the implicit obedience which husbandmen pay to the laws of their country, however adverse these laws may be to the interest of their order. If they had all those malevolent intentions which are imputed to them, how easy would it be for them and their families, who in the course of their business are perpetually exploring the fields, to crush the game in embryo, and in a great measure extinguish them!

When population was inconsiderable, and a great part of the country wild and desolate, savage animals were no doubt numerous, and would make injurious inroads on the cultivated parts of the country. In those barbarous times, it was the employment of the Barons and their retinues, in the intervals of their wars and feuds, to hunt

and

and destroy those animals. It was the most innocent
part of their employment, probably, not seldom the most
useful, and therefore might justly be considered as honourable. Posterity is perhaps more indebted to those
Barons for having extirpated wolves from the island, than
for the greatest part of their military exploits. Hunting,
or the sport of the field, has, therefore, from those days
downward, been considered as a gentlemanly employment; and every young man who wished to be regarded
as a gentleman, has thought it necessary to qualify himself for being a sportsman. But, happily, those times
are now over. Marks of the industry of man, preparatory of more important and successful improvements, are
seen every where through the country. By the effects of
these, the game is not only diminished in numbers, but
those which remain are become naturally more shy and
careful of their own preservation. Hence now, instead
of manly activity and courage, much piddling patience,
much low mechanic cunning, is requisite to success in
killing game; nor is the service done to society any longer an apology for the practice of pursuing game. Whereever the industry of man has been extended, the numbers of the game are too limited to be injurious, and even
the nature of the ferocious in some measure changed.
The fox is the only animal of prey which is accounted
game in this country. Among the numerous flocks of
sheep which feed around the Cheviot, abundance of foxes
are to be found. The fox is hunted there, but it is only
for amusement; for years elapse without an instance occurring of a single lamb being devoured by foxes in a
whole parish. But if the advantage of society is not now
promoted by the pursuit of game, it is, as has been already
shown, almost always in some degree injured. The sportsman can seldom continue his pursuit in a great part of
this

this county for a few hundreds of yards, without committing some petty trespass. The anxiety of those who, by the laws and customs of the country, are the exclusive proprietors of game, to secure the enjoyment of it, may be injurious in other respects. The following instance will exemplify this:—In a county not far distant, the soil of a great part of which is fertile and highly cultivated, the depredations committed on the crops by rooks, particularly on the wheat, during the winter, was found to be so enormous, that some years ago a meeting of the husbandmen was held, to consult on the means of preventing so great a loss, as well to themselves as to the public. The method agreed on was, to employ a person to lay down paste, in which arsenic was mixed, in proper places, for poisoning the rooks. This was done, and succeeded. The rooks ate the paste, and numbers of their dead bodies were seen strewed in the fields. The husbandmen, rejoicing at the success of their scheme, were in hopes that perseverance would rid them of their wasteful enemies. But accounts no sooner reached the fox-hunters of the county than they took alarm, lest the foxes, eating the poisoned rooks, should take the deleterious potion into their stomachs, and perish, and the amusement of fox-hunting be thus disappointed. A meeting of the justices of the peace was therefore called, and injunctions published, forbidding all persons, under severe penalties, to expose poisonous substances in the fields. Thus the rooks are again allowed to multiply and prey on the corn fields.

Since the circumstances which gave rise to the laws and customs relating to game are entirely changed, it would probably be proper that some change should be made in those laws and customs. But it may, perhaps, be thought that too much has been already said on the subject; and therefore it shall be left to the judgment of the public.

CONCLU-

CONCLUSION.

MEANS OF IMPROVEMENT, &c.

HAVING now, with all possible diligence, explored the county of Lanark in its present situation and circumstances, and more especially in such as bear any relation to the most substantial national resource, the agriculture of the country, we come finally to inquire into the possibility of making such improvements in this, as to correspond with the advancement of the other industrious arts and the growing population, and so to give some prospect of security to the continued independence and prosperity of the nation. Nor ought we to despair of this being accomplished, notwithstanding some circumstances in the climate and soil adverse to fertility, which have been already noticed. Though the climate is unsteady, its variations seldom go to destructive extremes. If Britain seldom enjoys the sun in unclouded brightness, its corn fields and pastures are rarely parched with his scorching rays; successive showers at intervals, generally not very distant, irrigating the earth, and keeping up the verdure of the season. If the warmth is not sufficient to forward the progress, and exalt the juices of the most delicate fruits, it seldom fails, when aided by the industry of man, to bring those which are absolutely necessary to the support of the inhabitants, to maturity. The soil too, though subdued with more difficulty than in countries more favoured by Nature, yields at length to the efforts of cultivation, as may be seen from numberless proofs over the country. Though agriculture, as a

regular

regular art, is but in its infancy, and has laboured under many difficulties, it has made considerable progress, and has paved the way for more important improvements. Even in these northern districts, instances occur of land, formerly barren, now producing corn crops, which would be thought large in countries naturally more fertile. When we consider also the spirit and enterprise of the inhabitants, and how far they have gone beyond the neighbouring nations in other branches of industry, there is no cause to think they should be deficient in the cultivation of the soil; and we may reasonably hope that, with proper encouragement, the agriculture of the country might not only furnish an ample support for the present inhabitants, but for a much more extensive population. Though speculative calculations are not greatly to be depended on, it may be agreeable to see that such might, in some measure, be the case, even in the populous and comparatively barren district which is the subject of this Report. The total surface extent has been supposed to be 445,440 acres. The cultivated lands, and wastes capable of being cultivated for corn, may be about the half of this extent, or 222,700. Agreeable to the practice of this and neighbouring counties, as stated page 81, of allowing the land which has been cultivated, to meliorate by pasturing, let us suppose two-thirds to be always in green crops, fallow, hay, and pasture. If 74,000 acres, the remaining third, were cultivated for the different kinds of bread-corn*, with all the attention which has been bestowed on particular spots, it does not seem to be extravagant to suppose, that each acre, on an average, might furnish bread for two people, besides seed and the food

---

* It is to be observed, that the poor use a great deal of meal made of barley and pease.

of

of farm-horses; and thus the produce of the county would feed all its present inhabitants, and furnish a part of what is consumed by other horses, &c. Nor need we limit the fertility of the county to this extent. Well cultivated land returned to pasture would still be increasing in fertility*; every increase in crop would occasion an increased quantity of manure; and the fertility of the land, and the quantity of corn it could produce, would gradually become greater.

A person

---

* The celebrated Mr. KIRWAN, in his Essay on Manures, &c. asserts, that pasture grounds are always diminishing in fertility, though in a less degree than those from which a crop is annually carried off. The experienced husbandman will be in little danger of giving implicit credit to this; but as so great an authority may stumble some, it may be proper to investigate the matter. By the late discoveries in chemistry, it has been proved that charcoal combined with pure air, exists in a state of gas, and that one hundredth part of the atmosphere consists of this combination. The same gas is perpetually escaping from the earth's surface, in the burning of combustibles, the fermentation of fermentable bodies, and the respiration of animals. Hence it might be supposed, that at length the pure air would be wholly contaminated, and no longer respirable. But this seeming evil, by the wise appointment of Providence, becomes an unspeakable good. The illustrious SENEBIER has prove by undeniable experiments, that the leaves of plants are endowed with the power of absorbing this gas, digesting it in their organs, restoring the pure air in sun-light, and retaining the charcoal, which contributes to increase their size. Thus pasture grasses, which abound in horizontal leaves, are perpetually drawing nourishment from the atmosphere, and all those leaves which are left to decay, add, from year to year, to the fertility of the soil. Besides, the ground that is covered with a close turf, is defended from the battering storms and washing rains; the chilling cold is also excluded, and the subterranean warmth retained; the effect of which is, to expand the soil, and make it mellow and friable towards the surface, without being open and porous. In this manner good pastures are always increasing in fertility, as experience has uniformly proved, and not diminishing, as Mr. KIRWAN supposes. When the accumulation of decayed herbage is turned over with the plough and mixed in the soil, its decomposition is facilitated, and it becomes a plentiful stock of direct
vegetable

A person who, from early life, has been taught to ad-
mire the superiority of the British constitution, as the
most favourable to civil liberty, will naturally be led to
attribute the national prosperity chiefly to this cause.
From a justifiable partiality to the institutions of his
country, he will view with pleasure the attention which
has been paid to the comfort and convenience of the
whole society, and the equitable laws which encompass
the rights and properties of the meanest. He will consider the administration of these as having called forth
the native energy of the people, and fostered that prosperity to which they are arrived. Convinced that the
perfection of political liberty is the happiness of a nation, it will scarcely be possible for such a one to avoid
turning his thoughts that way, when proposing measures for the improvement of agriculture; and surely
there can be little cause to fear, that, in so doing, he can
give offence to a public, whose partialities, from the same
cause, must be the same with his own. His fate, indeed,
may be like that of the prophetess Cassandra, or of those
boys, mentioned in an ancient book which we all pre-
tend to revere, who, " sitting in the market-place, piped,
but no man danced,—grieved, but no man mourned."
It is incumbent on him, however, to state what he ap-
prehends to be the truth. It may not be useful at present, but it will at least be innocent; and, should it furnish hints to be improved on in future times, it will not
be without effect.

In the prosecution of this inquiry into the means of

---

vegetable food, from which the succeeding tillage crops derive abundant support; and the texture of the soil being also improved, an increased produce of the grasses is obtained when the ground is again
laid to rest; for all pastures, in course of time, degenerate, except on
soils of the most favourable texture, lying on a dry bottom.

promoting

promoting the success of agriculture, we shall first consider how the great obstacles enumerated in the last chapter, may be removed, taking them in the order in which
they are stated, and then venture a few hints for improvements in some other particulars.

The first obstacle stated, was the low estimation in
which husbandmen are held in the scale of society, and,
more particularly, the marks of degradation affixed to
the profession, by certain laws and customs of the country. Mankind are so much governed by habit, that what
is perpetually passing under their eye scarcely strikes
them. For this reason, it is probable, this obstacle will
be thought frivolous; even its existence will, perhaps,
not be admitted. But let it be recollected, that roturier,
boor, clown, country fellow, &c. are terms of contempt
in all the countries of Europe: and no person of ordinary observation can avoid seeing, that the other classes
of society, in their intercourse with country people, look
as if they would say or think, " Rusticus es, Coridon !"
A girl bred in a neighbouring town, accompanied a party
on a visit to a country house : the country people entertained their city visitors with great hospitality. Tea was
served, accompanied with the richest cream the dairy
could afford. The girl, on her return, observed, that
every thing was coarse in the country : how clumsy,
said she, was the cream we got to tea ! The instances
given in former parts of this Report, of mill multures,
work at roads, customs at fairs, &c. may serve to illustrate the remainder of what is stated as an obstacle.

The great trust committed to husbandmen, of propagating and managing the provisions of the community, is
an office of the first importance, and is, surely, in the
eye of reason, entitled to respect. The difficulties, from
soil and climate, with which the husbandmen of these

CLYDESDALE.]   Q   northern

northern parts must struggle, are sufficiently discouraging. Instead of additional burthens, every thing should be done to encourage and enliven their useful toil. All laws and usages, therefore, which bear unequally hard upon husbandmen, or affix any thing like degradation or servitude to the profession,—all the privileges of other classes, which, in any manner, are at variance with their just rights, ought to be abolished ; and, so far as it were practicable, and consistent with justice to the rest of the society, every encouraging mark of distinction should be bestowed. The Board of Agriculture are now the established patrons of the art ; and to them its professors look up for relief and encouragement. To their wisdom these rude hints are submitted.

If the legal degradations alluded to (the more weighty part of the obstacle) were removed, the popular prejudices would follow of course; and to this the correct conduct of husbandmen would greatly contribute. There is no profession in which there are fewer temptations to fraud and low cunning than that of agriculture. No nefarious practice, which a husbandman, in the course of his business, could resort to, nor even a whole life of deceit, could add any thing considerable to his wealth ; but the discovery of his baseness would probably ruin him. For the proof of this, we may have recourse to common experience. Look to the thriving husbandman ! He is, indeed, active in his business and attentive to his interest ; but, in his transactions with others, he is open, candid, and sincere, ever studious to avoid taking, or appearing to take, undue advantage. Again, those who are guilty of knavish practices, feeling themselves depressed with a sense of the impropriety of their own conduct, and the imputations it draws upon them, are generally feeble in their exertions, and their affairs unprosperous. Neither does

does this line of business prompt to other unsocial passions. The interest of a husbandman is never hurt by the success of neighbours; on the contrary, a number of industrious and successful cultivators is the best neighbourhood for each of them. Hence there is little room for malice and envy arising among them. Those, therefore, who allow themselves to act a mean, shuffling, dishonest, or unhandsome part, are doubly culpable, deserve the abhorrence of all the virtuous of their order, and ought to be excluded from the benefit of the society, having forfeited all right to the character of husbandman. The Writer thinks himself thus justified in earnestly admonishing the people of this profession, as they regard their own interest, the honour and reputation of their order, and the success of the art in which they are engaged, that, while they are diligent in their labours at home, and attentive to all the minutiæ from which their interest and advantage is derived, when they are necessarily called out to the commerce of the world, they would avoid all deceit, all the chicanery of mean jobbers, and be open, just, and liberal, in all their transactions. Let strict probity be the rule of their conduct; in short, let them be what husbandmen really ought to be, and then they will deserve the respect of the public, and probably at length attain it.

The second obstacle, was the discordance between the landed and manufacturing interests. The consequences of this were exemplified in the operations of the corn laws. It requires no great ingenuity to discover, that the jealousies subsisting between these two classes are futile, their interests being so intimately connected, that whatsoever depresses the one, must be injurious to the other. The advancement of manufactures, and the increase of the numbers they can employ and pay, by en-

Q 2　　　　　　　　larging

larging the market for all kinds of land produce, is the
most powerful inducement to improvements in cultiva-
vation, and perhaps the only effectual one to attempt the
improvement of land originally barren. The greater the
success of the cultivation of the country in multiplying
provisions, the support of manufacturing people will be
the more certain and regular, and the bread made from
the corn produced in the country, incomparably more
wholesome than that from foreign corn, frequently heap-
ed up in granaries till the mephitic effluvia emitted from
it is insupportable.

There is no class of people so nearly interested as the
gentlemen of landed property, in the general prosperity
of the country; and the prosperity of a country must
depend on the numbers to whom its arts and manufactures can afford the means of subsistence. This county
furnishes a striking instance of the truth of these posi-
tions. The energy of its industry has, in the course of
ten years, occasioned an increase of population to the
amount of 22,000; and hence it is at present, perhaps,
the best market in Europe for all the ordinary productions of land. As land and its productions are valuable
in proportion to the numbers who can come to market
with money in their hands sufficient to purchase as much
of those productions as their necessities require, it does not
seem to be prudent to attempt increasing that value by
means of more doubtful consequence. The late act for
regulating the importation of corn, which has given
such great offence to the labouring poor, seems to be a
measure of this nature. If the agriculture of Britain
can supply its inhabitants with food, they will buy their
food from their countrymen, the British farmers, and
the greater number of them will give for that food all
they have to spare. But if the annual quantity of land
                                                    produce

produce does not keep pace with the increase of population, why should not the deficiency be supplied at the rate at which corn can be imported from other countries? Opposing this, seems to be setting bounds to the population and prosperity of the country : for though the energetic industry of the British people enables them to purchase provisions at a rate somewhat higher than that at which the consumers of the neighbouring countries obtain them, yet it may be dangerous to raise that difference still higher by forced measures. The spring, already highly strained by enormous taxes on consumable commodities, and other circumstances which inflame the expense of living, may at length be broken ; and then, farewell to the prosperity of the country, and with it, to the high value of land ! ! ! Let not landholders, then, for the present purpose of raising or supporting a high rent-roll, embrace a measure which may endanger the future interest and dignity of them and their posterity. The rent of land has, of late, been rapidly advancing to an extent which may satisfy the avidity of the most sanguine. If the state of the country can support it, it will continue ; if not, it ought to be lowered : for it will be in vain to attempt supporting it by laying the labouring poor under contributions, which they are unable to bear. Renounce, therefore, O British landholders ! the pedlar-like expedient of monopoly, which has been so often and so justly reprobated in the conduct of the mercantile and manufacturing classes. When they solicit monopolies, they proceed upon well-founded motives. Under the pretence of public welfare, they have their own interest full in view, and mean to enrich themselves at the expense of other branches of society. But your interest and happiness are inseparably connected with those of the industrious poor. They are

the

the bees who collect honey from every flower which blossoms on your fields, to be ultimately deposited in your cells. If they are oppressed, ye suffer; if they cannot support themselves, ye must feed them. It is equally your duty and interest not to burthen, but to cherish and protect that class; to promote their comforts; to watch over their morals, the surest pledge of their usefulness; and, in short, to give their spirited industry unlimited scope.

But the law, by which a bounty is offered to encourage the exportation of British corn to foreign countries, has been supposed to be the cause of rendering the quantity more abundant in the home market, and so to have lowered, instead of raising the price of corn. Whatever has been the cause, this law seems not only to be unnecessary, but unjust. If it has had a tendency to raise the price of corn in this country above that in other countries, it was doing injustice to the industrious consumers, and checking the progress of manufacture. If it lowered the price, it was defrauding the cultivator of a part of his just recompense, which ought always to be in some proportion to the vigour and success of his industry. At any rate, it was abstracting a part of the funds levied for the support of Government, from the whole community, to favour one part of it in opposition to another, or rather, perhaps, to favour other nations in opposition to the whole. The occasional bounties given to encourage the importation of corn, on the prospect of scarcity, seem to be no less partial and improper. In a nation where commerce is so flourishing and extensive, and where all kinds of industrious labour are so liberally rewarded, there can be little cause to fear but corn, as long as it is to be found in the world, will be brought to so good a market. It would, no doubt, be somewhat

higher

higher priced in proportion to the scarcity; but, in justice to the husbandman, the price ought to be something better, when the quantity raised is less.

Had the commerce in corn been always free, secure from the interference of executive government, and without restraints or bounties of any kind, it is probable that agriculture would not have been left so far behind. But in the state in which it now stands, compared with other industrious arts, it might, perhaps, be of importance, that some small duty were laid on foreign corn imported, to compensate the husbandman, in some measure, for the difficulties of soil and climate, high rents, great expense of cultivation, &c. The great proficiency made in the different branches of manufacture, by the expertness of artisans, and the abridgement of labour, from the invention and improvements of machinery, may enable manufacturing people to give their brethren, the cultivators of the country, without any disadvantage, this indulgence: and, if agriculture should be thus enabled to provide amply for the wants of all the inhabitants, it would, in the end, tend greatly to the advantage of manufacture. Such a duty, however, ought to be very moderate, and not fluctuating with the market prices, but the same at all times. The Legislature might also assist the agriculture of the country to correspond more completely with an increasing population, by giving all necessary indulgence and encouragement to the cultivation of every kind of improveable wastes. This it would be the more enabled to do, if the money bestowed in bounties on exportation and importation were saved. A small duty levied on all corn imported, and committed to the management of the Board of Agriculture, would form a fund, which, under the direction of that well-informed body, might tend greatly to promote the

Q 4

fertility

fertility of the country. Besides amendments which might be made in the cultivation of the lands already under culture, there are still many improveable wastes lying untouched : in particular, a moderate encourage-ment, applied to promote the improvement of the ex-tensive peat-mosses of this county, might bring those useless and ugly wastes to be valuable fields.

It would tend to destroy the unfortunate alienation between the agricultural and manufacturing classes, if the latter, instead of being collected in groups, were more regularly dispersed over the country. It is now obvious, that, so far from being necessary to have all the operative people employed in the principal manufactures carried on in these provinces, closely assembled together, it is frequently more convenient that they should be sepa-rate. It has long been the custom of the great manufac-turers of Glasgow and Paisley, to employ weavers, &c. living 40 or 50 miles distant, as well as in intermediate places ; and, not many years ago, a considerable part of the linen manufacture of Edinburgh was executed by weavers of this county. If any disadvantage had been experienced from this practice, it would have been dis-continued, and the operative people invited to assemble more closely : on the contrary, it would appear to be, in every respect, more advantageous. Operative people, living separately in the country, are generally more orderly and attentive to their business, and less given to cabal and riot. They save something by the difference of house-rent, carriage of fuel, provisions, &c. and are sometimes enabled to provide a little for times of dis-ease and old age. To the general cause of agriculture it would be very advantageous. The offals of all kinds, which amass around the habitations of man, though nau-seous when accumulated, are, by the wise appointment
of

of Providence, highly beneficial, when spread over the earth. These can only be carried to a short distance from great towns, and much is washed away and lost. But as every thing cast off by the inhabitants of the country produces fertility, the effect must be greater in proportion to the numbers and regular dispersion of the inhabitants. Connexion would be formed between husbandmen and operative manufacturers residing among them, and a mutual interchange of good offices would take place. While the former accommodated the latter in carriages and the like, the latter, with their families, would be disposed, in return, to assist in some of the most urgent labours of the field. Habituated to these occasional exercises, they would be better fitted, and take more pleasure in performing them. Some of the children of such manufacturing families would incline to work always in the fields, and, as they grew up, engage in the labours of agriculture; and thus manufacture, instead of abstracting the labourers from agriculture, would serve to recruit them. On these considerations it would appear to be proper, that convenient houses for the reception of weavers, or such other operators in manufacture as perform their work single-handed, were erected, one or more on a farm, on all farms which lay near the high roads communicating with such towns as are the centres of their manufacture. Each of these houses should have as much land attached to it as would serve to feed a milch cow, and furnish potatoes and greens to the family, to be cultivated by way of relaxation in spare hours. This topic is trite; but the advantages which would arise from its being put in execution, are too numerous for us to take time to expatiate on them at present, and too obvious to require it.

The third obstacle mentioned was, the overstretched land-

land-rents, &c. It is difficult to say how this obstacle can be removed. While farmers are to be found disposed to take leases of land at an extravagant rent, it would be talking to the winds to attempt to persuade landholders to rent their lands lower than the value which, according to the partial eye with which every man regards that which is his own, they think it worth; and all other means to regulate the rent of land would be an encroachment on property, the protection of which every British subject thinks his birth-right. This having been the basis of our prosperity, it is the earnest wish of the Writer it may never be infringed. The adjustment of rents must, therefore, be left to time, and the liberality and good sense of the landlords.

The same may be said of the fourth obstacle, the deficiency of the capital employed in agriculture for the purpose of carrying the improvement of the country to its full extent. While the rent of land continues to rise as fast, or perhaps faster than its value increases, the stock engaged in agriculture cannot much augment; and while greater profits, and less laborious exertions, await other branches of industry, little capital can be drawn from thence to agriculture. It is true, indeed, that people frequently purchase lands with part of the surplus wealth amassed in trade, and apply part to the cultivation; but the quantity of food for the support of industrious consumers, raised by such cultivators, is always far short of that which the frugal economy of professional husbandmen produces. Besides, the recoil of wealth from trade to the purchase and cultivation of land, is limited, all lands held under entail being out of its reach; and such lands generally stand most in need of farther improvement.

The fifth obstacle stated (and, perhaps, the greatest to the

the general improvement of agriculture), was the great land properties, the shortness of the leases given to the farmers who occupy them, and the frequency of changing those farmers at the end of every lease. It was already observed that, in consequence of this, the interests of the landlord and farmer are at perpetual variance. Those who have the management of great estates are sensible that it is so, and many expedients have been devised to prevent the farmer from getting the advantage. Pretended doctors in agriculture have been consulted, by whom rules for cultivation, and other profound schemes, have been prescribed, to bind farmers to promote the advantage of their landlords, without regarding their own. But is it rational to expect energy from such farmers, who must ever bear in their minds the lamentation of the poet?

" Sic vos, non vobis, fertis aratra, boves !"

It is sufficient barely to state this system, to show how unfavourable it must be to the success of agriculture. It is very questionable, if a continued perseverance in it would be consistent with the private interest of the landholder. Regulations are prescribed, and frequently forgotten; but no attention which can be paid by the managers of great estates, can make farmers strictly fulfil the letter of them, when they apprehend such regulations are disadvantageous to themselves; nor have the decisions of the courts of justice been favourable to the enforcing of such capricious restrictions. The chief effect of those restrictions, therefore, is either to damp the spirit of farmers by enforcing them, or to injure their morals by the temptations still found to elude them. Hence it would appear, that the private restraints imposed

posed by landholders on their farmers, with a view to raise the rents against the end of the first leases, have as little efficacy as the public restraints on the commerce of corn. The rent, or yearly value of land, can be increased only in three ways: 1*st*, when the growing prosperity of the country occasions a greater abundance of money, and, of course, the value of money sinks, the money rent of land naturally rises in the same proportion; 2*dly*, proprietors of land, by laying out expense on the improvement of it, such as, for enclosing, draining, or the purchase of manure, may increase the value; 3*dly*, an industrious farmer may, in the course of a lease, increase the fertility, and, consequently, the yearly value of the farm. All increase of rent, which is occasioned in the first way, evidently belongs solely to the proprietor. The increase, in the second way, being purchased at his expense, equally does; and, accordingly, both proprietor and farmer have those improvements in view in fixing the rent. The increase of rent, occasioned the third way, seems, in equity, to belong chiefly to the farmer; that is, so far as the increase of fertility has been really occasioned by his skill and attention: for the first cause of the increase of rent, and even the second, in the article of enclosing, &c. is so intimately combined with the third, that it is difficult precisely to determine the amount of each. However, it would seem that a good farmer is, in equity, entitled to some recompense for his uncommon toil, skill, and attention, more than he can obtain in the course of an ordinary lease. At least, there is no doubt that all farmers think so; and, while their landlords are of a different opinion, and disposed, at the end of every lease, to lett farms to new tenants, for a small advance of rent above what the old ones think themselves able to pay, farmers will always

be

be cautious that landlords shall obtain as little advantage as possible from their labours.

In order to get rid of this great bar to the progress of agriculture, it would seem proper, that the possession of farmers were made perpetual. It is probable, that landholders may stare at such a proposition ; but it must be remembered, we are not addressing ourselves to the prejudices of any particular class, but considering what might contribute most to the general good. Besides, no Agrarian law is meant here to be proposed, nor any infringement on the established rights of individuals. Though the possession of farmers were rendered perpetual, or to last so long as the same family chose to continue in the farm, the landlord would still have the same right to the regular payment of rent, and to eject such farmers as should be found deficient in that duty. In order to adjust the quantum of rent, according to the circumstances of the country, and the state of the farm, at the end of every 20 years an inquest of impartial men of judgment might be appointed, to take cognizance of such alterations as might have taken place during that period, and affixing the rent for the next 20 years accordingly, with an equitable regard to the just rights of both parties ; and this might frequently be done by the parties themselves, without any assistance. The farmer should have the property of the trees he might plant in places unfit for tillage. If it should be no longer convenient for the family of a farmer to continue in the possession, the intention of removing, with the causes which induced it, should be intimated, in proper time, to the landlord, for his consideration, that, if he should think it reasonable to accept of the resignation, he might appoint the farm to be visited, and, in case it had suffered

any

any damage, this damage should be estimated, and, like rent, should be a preferable debt.

In this manner the interest of the landholders would, probably, be as effectually secured as by all the complex refinements which have been devised for that purpose. They, and their people of business, would, at the same time, be freed from the harassing circumstances accompanying the attempts to carry those refinements into execution. To enjoy the income of a great estate, and to produce that income by cultivation, are two employments of a very different nature. It is probable, that those whom Providence has destined to the first, may frequently be mistaken in many things which relate to the last. It would, perhaps, be better therefore, to leave the cultivation of the country to those who are engaged in it. A landlord would have no cause to fear that, by making his farmers somewhat more respectable, he would be less so. He would not be less the object of respect, that he was no longer the object of dread ; and, surely, the respect of independent men is more valuable than that of slaves.

We should thus have a permanent race of husbandmen, each generation of whom would derive knowledge in their profession from the experience of their predecessors. Freed from embarrassing restraints, and satisfied that they and their children would enjoy the just recompense of their important labours, they would be studious to know, and diligent to execute, whatsoever might tend to increase the fertility of their farms. Though the increase of the capital employed in agriculture might be slow, it would be more certain among a steady race of husbandmen, than when they are perpetually shifting ; and, as it increased, the operations of agriculture would be more energetic ;

energetic; and the high rents, and any reasonable addition which might be made at the end of 20 years, would be less distressing, as the operations of agriculture became more powerful.

Before we quit this part of the subject, it is proper to observe, that, though it is recommended to landholders, for the sake of their own ease and advantage, as well as for the good of the public, to give farmers a more steady unencumbered tenure of their possessions, it is not meant that those landholders should be deprived of the choice of using their property as they think proper; but there could, surely, be no harm in empowering incumbents on entailed estates, if they should choose, to give perpetual leases to farmers, when such leases had no tendency to injure the rental: nor is it wished that such land proprietors as incline to make improvements on any part of their estates, should renounce that respectable amusement. They cannot, perhaps, entertain themselves in a manner more rational or more useful. Their greater leisure for inquiry, and their superior ability of making experiments, might lead to discoveries, important to the cause of agriculture, which could not so certainly be derived from the practice of ordinary husbandmen. Happily, the late discoveries in chemistry and natural history, have now overthrown those false and absurd theories, with which the practice of agriculture has too long been encumbered. Speculative people, who have aimed at being more wise and learned than the laws which Nature prescribes to us, seem to admit, and, to take an easier way of attainment than she has directed, have, one after another, advanced their systems. By those blind guides, the agriculturist has been led to grope and stumble in the dark, and, when he was made to believe he had laid hold of the truth, has found, in the end, that he only grasped error.

error. The discoveries alluded to have restored him to the day, and taught him not to enter the cavern of inquiry without the torch of experiment in his hand. Though those discoveries should proceed no farther, they have been a great benefit to agriculture; but, it is probable, they will still have more considerable effect. A number of ingenious and benevolent men are generously employed in making researches into the secrets of Nature, for the benefit of mankind, whose philanthropy, compared with the sordid pursuits in which the bulk of the world are engaged, does honour to humanity. In the course of their investigations on the different properties of matter, new sources of fertility may be discovered, and new means of increasing our harvests, at a moderate expense, may be found. It would be an employment becoming landed gentlemen to encourage such researches, and to carry experiments into practice, on such discoveries as may result, in order to make the utility evident to husbandmen.

If this principle be admitted, that the more those who are employed in agriculture are engaged, by the prospect of advancing their own interest, to conduct it with vigour, the improvement of the country will be more successfully promoted, little need be said respecting the sixth obstacle. From 100 to 150 acres of arable land is sufficient employment for the activity of one man. If the farmhouse is placed nearly in the centre, manure can be carried, work conducted, or any thing looked after, on all parts, with very little loss of time; and, on farms much more extensive, it is obvious this cannot be the case. To repeat more arguments in favour of moderate farms, seems now to be needless.

The seventh obstacle, the effect of the game laws, it has been already observed, is of an inferior kind, and the

nature

nature and extent of the injuries it occasions, have been fully illustrated. We therefore proceed to offer a few hints on other particulars, by which the improvement of agriculture might be promoted.

We shall begin with the natural pastures; that is to say, those parts of the country which, from their elevation in the atmosphere, or the inequalities of their surface, are unfit for tillage and corn crops. All thoughts of making any improvements on those, having been long ago given up, they are left to the sheep, to gather what sustenance from them they may naturally yield; and a few sheep farmers have engrossed the possession of a large extent of country. In this condition, those pastures which, in this county, are about half the surface extent, must be gradually diminishing in their real value, and the subsistence of the flocks, when deprived of the aid of human industry, more scanty and precarious, and, of course, in proportion to the hardships to which the animals are exposed, the benefit derived from their wool and mutton must be less considerable. The injuries which time operates on a neglected country being slow, are less palpable, and, perhaps, will not be generally admitted; but a little reflection will convince the unprejudiced mind, that they must certainly take place. Every new gully which a torrent forms, must diminish the extent of surface pasture. Wherever surface water is, by any accident, detained, it will prey upon the esculent grasses, and nourish mosses, and other useless herbage, in their place. The herbage rejected by sheep will gradually prevail over that which is eaten; and thus the quantity of pasture must be continually diminishing. It would therefore seem proper, that sheep were put more under the protection of man, by engaging a greater number of people to follow a pastoral life; so that each, in

CLYDESDALE.]   R   the

the pursuit of his own interest, by providing better accommodation for his flocks, may render the pasture
grounds of the country more valuable to the public. The
first step to this purpose would be, to begin with making
a sufficient number of enclosures on the pastures: for
this the hills always furnish abundance of stones; and the
practice of the inhabitants of the district of Carrick, in
Ayrshire, and of the counties of Wigton, Kirkcudbright,
and the western part of Dumfries, is a very good example.
But the Author of this Report, in a former work, to
which, it is hoped, the public will pardon his repeatedly
referring, has, from inferences drawn from the nature of
sheep, stated a number of observations on the means of
deriving the greatest advantage from pasturing them. As
it would be improper to swell this Report with long extracts, he begs leave again to refer to that work*. It is
true, indeed, the experience from which those observations were derived was narrow; but he having since, in
an extensive tour made through the southern pastoral
district of Scotland, &c. in the year 1793, under the
direction of the British Wool Society†, found nothing
material to contradict what he had before advanced, with
the greater confidence recommends it to the public.

If some such practice as has been there recommended,
were to be adopted, the pastures, instead of degenerating
by neglect, would be gradually improved. The flocks,
under the more immediate protection of a number of
industrious shepherds, employed in supplying their occasional wants, and defending them from incidental injuries, would suffer less by disease and hardships, and the

* NAISMITH's Thoughts on Industry, &c. book iv. chapters 1, 2, 3,
and 4.
† See NAISMITH's Tour, &c. Edin. printed by W. SMELLIE, 1795.

profit

profit arising from tending them would be much more considerable. Sheep attended in this manner would always be in good habit, and the owners could thus dispose, at all times, of a great part of their surplus stock immediately to the butcher. Being thus freed from the necessity of adapting their rule of breeding to the prejudices of a certain set of purchasers, they would naturally be led to consider what kind of stock, taking carcasses and wool, and all circumstances together, would be most advantageous in their respective situations. And as any addition to the value in quantity or quality of the wool, not counterbalanced by other circumstances, is a valuable consideration, they would be induced to pursue every prudent measure to improve the fleece. What the patriotic intentions of the British Wool Society, from a concurrence of adverse contingencies, has not been able to effect, would thus take place of itself.

A number of collateral illustrations might here be given, in support of what is above stated; but all of them, upon examination, seem so palpable, that to trouble the intelligent reader with the perusal of them would be offering some kind of affront to his understanding. We go on, therefore, in the next place, to observe, that it would be advantageous to the success of agriculture, if husbandmen, as a particular class of the general society, were more intimately united. The insulated situation and sequestered manner necessarily attached to the life of husbandmen, has been already mentioned, as tending, in some degree, to depress that order; and such measures as they have taken for their joint support have been described. But other important purposes might be effectuated by a complete association of all husbandmen into particular societies, consisting of moderate contiguous districts, so as to be most convenient for the members

R 2   assembling

assembling or communicating on proper occasions. All
the members of each of those societies ought to be bound
to submit to the general will of the society, and to obey
its laws. Each member should also pay stated contributions into a common purse, to be applied, as circumstances require, for the benefit of the society. Com-
mittees should be annually chosen to manage the business, and construct such laws as may be requisite for the
common good; and these might afterwards be submitted
to the general meetings for their sanction. From the
wisdom of the regulations, and integrity of administration, observable among the Friendly Societies of artificers
and others, there is no cause to think that husbandmen
would, in any respect, be more deficient in the conduct
of public business, and especially such as was intimately
connected with their own profession. We may reasonably hope, therefore, that the operations of such an
association of husbandmen would, in many ways, con-
duce to promote the success of agriculture. It will
suffice to give the two following instances, by way of
illustration.

1st, It might be made a general law in every society,
that all the members were to be bound to extirpate,
while in the flower, all those weeds, within their respective possessions, which bear seeds in any degree winged,
or transportable by the winds, such as the common dock
(rumex vulgaris), the burr thistle (carduus lanceolatus),
the marsh thistle (carduus palustris), the prickly thistle
(serratula arvensis), the sow thistle (sonchus oleraceus),
hawkweed (heracium pilosella), the dandelion (leontodon
taraxacum), tussilagow (farfarum), ragwort and ground-
sel (senecio jacobea et vulgaris), &c. whether these were
growing in the pastures and cultivated lands, or by the
sides of highways, brooks, ditches, or fences, any where
within

within the bounds of the farm. That this work might be duly performed, one or more of the members of each district might, by turns, be appointed censors, to visit all the farms within it at the proper seasons, and report. If any individuals had been neglectful, fines, proportioned to the degree of negligence, ought to be imposed and levied by the managers of the society, for the use of the common stock. In this manner, those weeds which are so great a nuisance, by being collected before the seed was formed, might be converted into useful manure; and when the perpetual influx of adventitious seeds was checked, each husbandman would have only to war against the perennial roots of such as were within his own farm, and would more easily subdue them.

2dly, The common stock of every society might be chiefly employed for the protection and defence of the property of all the members within the district, against all the different kinds of enemies to which the produce of the fields is exposed; and that not only by checking actual depredations, but by preventing the repetition of them. Thus, if the attention of a number of societies, spread throughout the country, and possessed of proper funds, were continually employed in counteracting the injuries done to the property of husbandmen by the sparrow tribe, by rooks, by rats, by moles, &c. the immediate depredations of these numerous wasteful enemies would not only be controuled, but, by a system of hostility regularly pursued against them, the different races would be so far subdued, as soon to become no longer formidable. The same would be the case with all the idle breakers of fences, and wanton destroyers of the fruits of the field. So soon as a few examples had convinced them that they were no longer committing trespasses on the property of a forlorn individual, but on

some

some member of a watchful society, ready to take the alarm, and in condition to seek redress, they would be much more cautious to avoid offending; and thus the prevention of crimes, which is still of more importance to society than the punishing of them, would, in some measure, take place.

Agriculture would derive many other advantages from husbandmen thus acting collectively. " As iron sharpen- " eth iron, so doth a man the countenance of his friend." In their frequent communications, their transactions in the line of their profession would be a common topic of conversation. Every one would be better informed of what was going on among his neighbours; and an emu- lation would take place among all, to vie with those who were most active and expert. If any superior practices in agriculture had taken place in another district, such asso- ciations would be well adapted to procure information concerning them; and, by a judicious application of part of the common funds, to ascertain, by experiments, how far such practices were suited to the district in ques- tion. By the same means, experiments might be made with the seeds of plants not in common use, to know in what degree they would tend to augment the general produce. For these purposes a small piece of experi- ment ground, in each district, might be convenient.

From the general tenor of this Report, by which it has been recommended to make it the interest of those who are employed in the cultivation of the country, to carry it to the greatest possible extent, and to confide principally on them for that purpose, it cannot be ex- pected that much will be said on the practice of husband- men individually. In the former Report it was men- tioned, that the lateness of the Lanarkshire harvests was one great cause why they were less productive. The

cause

cause of the late harvests was ascribed to a combination
of the effects of soil and climate, minutely stated in the
first Chapter of this Work; and two things were recommended to palliate this evil: 1*st*, in the cultivation of
the land, to attend to every thing proper to put it in
such condition, that the seed, when sown upon it, may
enjoy every advantage which the weather offers; and
this is all that can be done to forward the harvest: 2*dly*,
diligently to pursue every possible means of getting the
corns quickly and safely gathered in, which is all that
can be done to render the produce of the soil effec-
tive.

With respect to the first, dividing the particles of the
soil, in a certain degree, so that while it gives admission
to the roots of the plants committed to it, it may retain
a sufficient quantity of moisture for their support during
the summer's droughts, is requisite. Both labour and
manure are applied for this purpose. But, to effect this,
labour can only be applied in a dry season; and, where
the sub-stratum is impermeable, and the soil dense, by
the first great rains which succeed, the effects of labour
are, in a great measure, undone. It is presumed that all
practical husbandmen will acknowledge this to be true;
and therefore, since the expense of labour is now become
so excessive, it would seem to be an object of great importance, to consider by what means the least possible
application of it might have the greatest effect. It is
universally acknowledged, that the less every kind of
soil is overloaded with stagnant moisture, in wet times,
it is more disposed to retain a sufficient portion in
droughts; and that the influence of the weather, when
not prevented by the stagnation of water, tends to separate the parts of the soil, and prepare it to give admission to the roots of vegetables. The natural inference

R 4

from

from this is, that the greatest attention ought, at all times, to be paid to surface-draining, and laying the land in such a way, as quickly to throw off superfluous water, and to plough as much as convenient of what is intended for spring crops, with a deep furrow, in the early part of winter, that there may be sufficient time for the action of the weather to prepare it for the reception of the seed. It is true, indeed, that this operation of the weather, which prepares the soil for the intended crop, also fits it for cherishing the perennial roots of any weeds which may be lodged in it ; but every expert husbandman, in the course of his economy, will be attentive to subdue these : and, by proper alternations of pasture and tillage, of rotations between culmiferous and broad-leaved crops ; of hoeing and weeding, as far as circumstances admit, he will generally succeed ; but if he should not, recourse must be had to summer-fallow. In all other cases, except that of destroying the roots of weeds, it would probably be prudent to dispense with repeated ploughings for the same crop, and rather to solicit the assistance of natural agents, to concur with the operations of a less expensive labour. Manure, as well as labour, is applied for adapting the soil to the nourishment of useful vegetables. No husbandman entertains a doubt of the good effects of applying those substances which he has been taught to consider as manure, as he has never found them to fail, except when counteracted by labouring in improper seasons. After what has been said in the fifth Chapter of this Report, of the substances used for manure in this county, it is needless to attempt a classification of those substances. A collection of animal and vegetable matters will always be the most considerable, and, perhaps, the most important ; and their tendency to putrefaction is evidently the cause of

their

their beneficial effects. From the time they are committed to the ground till their total consumption, they
are perpetually aiding the operations of the weather
above-mentioned. It is therefore the business of husbandmen, by every means in their power, to increase
the quantity of this collection ; but it does not appear to
be necessary that they should be solicitous with respect
to the precise state in which it is applied : the state of the
weather, of the ground, the cessation of other business,
&c. will be the best rule. It is true, that, notwithstanding that the new lights which modern philosophy has
communicated, have taught us

" To know how little can be known,"

the itch for theorizing is so prevalent, that speculative
gentlemen are disposed to take the task out of the hand
of Nature, the most able operator, and advise husbandmen to elaborate, for the nourishment of plants, a certain degree of putrefaction proper for their food. This
seems to be unnecessary. The Writer of this has repeatedly had occasion, after the magazine of dung collected through the year was exhausted, to lay the dung
of cattle, and their litter recently made, on land prepared for wheat, in the month of September; and,
having marked the places where the unfermented dung
was laid, though the quantity was less in the proportion
it was less decomposed, he never could discover, through
the progress of the season, that the crop was inferior.
But though little delicacy need to be observed with respect to the time and state of applying manure, every
husbandman must have observed, that the more accurately it is spread, and the nearer it is kept to the surface, its efficacy is the greater. This naturally leads us
to

to attend to what is called top-dressing. The instances
are few, in which top-dressings can be advantageously
applied to the growing crops of corn in the northern
parts of Britain; but there is no place where they may
not be successfully applied to pasture and meadow lands;
and it deserves to be particularly noted, that many substances which have hitherto been found of little avail on
grounds which are in tillage, have proved very beneficial
to those which are in grass. It is believed, that every
kind of fossil which is disposed to crumble by the influence of the weather, will, after being thus decomposed,
improve the verdure on grass grounds; and it is certain,
that every addition which can be made to the quantity of
esculent herbage, on any ground, is the cause of that
ground being more fertile in future.

As to the second article, the in-gathering of the harvest, though not less important than the preparation of
the ground, less need be said concerning it, the husbandmen of this county being generally pretty active in
endeavouring to save the corn in rainy weather. One
practice, however, which has been successfully followed,
and ought to be more general, shall here be shortly
stated. When the corn is ready for the sickle, and the
weather so damp and rainy that it cannot be got reaped
sufficiently dry, the band of the sheaf is loosely tied
round, near the top, and the root end being properly
spread out, each sheaf, singly, is made to stand upon it.
This is here called *gaiting*. These must be carefully
attended to, and kept always standing, and they will
become dry, in the first interval of fair weather, in a
few hours. In the mean time, a sufficient number of
little hollow cones, formed of spars and branches of trees,
ought to be got ready on the corn-field. The first time
the corn is tolerably dry, it should be bound up, and
built.

built round these in little ricks, care being taken that, by using part of the sticks and branches, the lowest tier of sheaves may be kept from the damp ground, and the air allowed to pass under them. Such ricks may be built by a man standing on a short ladder, or portable scaffold, without compressing the sheaves by trampling on them; and if they are properly built, and the tops secured with a little thatch and a few ropes, they will stand safely, without the corn either springing or heating, till it is fully cured to be carried to the barn, or built in larger stacks. This will be thought laborious; but, in seasons when it is necessary, it is nothing to the labour which must be done, and the waste which ensues, after the corn has been drenched with repeated rains, and must be spread on the ground to dry.

It has been repeatedly hinted, that the future fertility of pasture ground is augmented in proportion to the additional quantity of sweet esculent herbage which can, by any means, be made to grow upon it; and for this purpose, all kinds of top-dressings have been recommended : but the Writer of this cannot help regretting, that so little attention has been hitherto paid to the propagating of our native grasses. As they are hardy and congenial to the country, the reproduction, from a well-chosen mixture of them, must certainly be greater and more unfailing, and tend more to lengthen the verdure of the year. White clover, which is the most universal, and, perhaps, the most important, is almost the only one which has been a subject of cultivation; but there are others, a little attention to the propagation of which would thicken the turf, and increase the quantity of esculent herbage. It is presumed, that no person who has paid attention to the different grasses prevalent in the fields, has failed to observe, that those pastures which

are

are found to be of the richest and best feeding kind, always abound, particularly, with the foxtail-grass, &c.; but it is needless to enumerate the sweet native grasses here, the Author, in the Appendix to a work to which he has frequently alluded, having described many of the most important, both for hay and pasture. About the same time, a much more able botanist, Mr. CURTIS, of London, was employed pretty much in the same way, and published a pamphlet, describing and recommending a good many of the native grasses, to which those who wish for better information may have recourse. It would, surely, not be difficult to propagate those grasses. It may always be observed, that any of them, of which a few are lightly interspersed through a field when first laid in pasture, from their superior congeniality to the soil and climate, annually extend, till the surface is in some measure replenished : it is probable, therefore, that if they were once more generally diffused, by mixing a small quantity of the seeds of any of them which might be preferred, with artificial grass-seeds, so many of their seeds would be shed on the ground, in the course of pasturage, that a rotation of tillage would not fully extirpate them. But, though it should not be safe to depend on this, if it were once common, it would be as easy to save the seeds of native grasses as those of rye-grass. By thus attending to every means of extending the propagation of esculent grasses, the pasture would be more valuable, and the alternations of pasture and tillage, practised in this and other counties, would, in the most certain and easy manner, ensure a continued increase of fertility in the produce of corn.

THE END.

Printed in the United Kingdom
by Lightning Source UK Ltd.
9677800001B/288